Experiments in Digital Fundamentals

A Systems Approach

David M. Buchla

Douglas Joksch

PEARSON

Boston   Columbus   Indianapolis   New York   San Francisco   Upper Saddle River
Amsterdam   Cape Town   Dubai   London   Madrid   Milan   Munich   Paris   Montreal   Toronto
Delhi   Mexico City   Sao Paulo   Sydney   Hong Kong   Seoul   Singapore   Taipei   Tokyo

Editor in Chief: Vernon R. Anthony
Senior Acquisitions Editor: Lindsey Prudhomme
Development Editor: Dan Trudden
Editorial Assistant: Yvette Schlarman
Director of Marketing: David Gesell
Marketing Manager: Harper Coles
Senior Marketing Coordinator: Alicia Wozniak
Senior Marketing Assistant: Les Roberts
Senior Managing Editor: JoEllen Gohr
Senior Project Manager: Rex Davidson
Senior Operations Supervisor: Pat Tonneman
Cover Image: Kayros Studio "Be Happy!"/Shutterstock.com
Media Project Manager: Karen Bretz
Composition: Naomi Sysak
Printer/Binder: Edwards Brothers Malloy
Cover Printer: Edwards Brothers Malloy

LabVIEW, Multisim, NI, Ultiboard, and National Instruments are trademarks and trade names of National Instruments. Other product and company names are trademarks or trade names of their respective companies.

10 9 8 7 6 5 4 3 2 1

ISBN-13: 978-0-13-298984-8
ISBN-10:   0-13-298984-0

Contents

iv

Cross-Reference to Floyd's *Digital Fundamentals: A Systems Approach*

Floyd, *Digital Fundamentals*	Experiment Number and Title
Chapter 1: Introduction to Digital Systems	1 Laboratory Instrument Familiarization 2 Constructing a Logic Probe
Chapter 2: Number Systems, Operations, and Codes	3 Number Systems
Chapter 3: Logic Gates and Gate Combinations	4 Logic Gates 5 More Logic Gates 6 Interpreting Manufacturer's Data Sheets
Chapter 4: Combinational Logic	7 Boolean Laws and DeMorgan's Theorems 8 Logic Circuit Simplification
Chapter 5: Functions of Combinational Logic	9 The Perfect Pencil Machine 10 The Molasses Tank 11 Adder and Magnitude Comparator 12 Combinational Logic Using Multiplexers 13 Combinational Logic Using DeMultiplexers
Chapter 6: Latches, Flip-Flops, and Timers	14 The D Latch and D Flip-Flop 15 The Fallen-Carton Detector 16 The J-K Flip-Flop 17 One-Shots and Astable Multivibrators
Chapter 7: Shift Registers	18 Shift Register Counters 19 Application of Shift Register Circuits 20 The Baseball Scoreboard
Chapter 8: Counters	21 Asynchronous Counters 22 Analysis of Counters with Decoding 23 Design of Synchronous Counters 24 The Traffic Signal Controller
Chapter 9: Programmable Logic	Introduction to Quartus II Software (On WebSite as Tutorial)
Chapter 10: Memory and Storage	25 Semiconductor Memories
Chapter 11: Data Transmission	26 Serial-to-Parallel Data Converter
Chapter 12: Signal Interfacing and Processing	27 D/A and A/D Conversion
Chapter 13: Data Processing and Control	28 Introduction to the Intel Processors 29 Application of Bus Systems

Preface

This laboratory manual has 29 experiments that follow the sequence of material in *Digital Fundamentals: A Systems Approach*, by Thomas L. Floyd. Electronics is traditionally a laboratory course that is meaningful to students only when theory is reinforced and observed in practice. To perform the experiments in this manual, each laboratory station should have a dual variable regulated power supply, a function generator, a multimeter, and a dual-channel oscilloscope. It is useful if the laboratory has a logic analyzer, but not required.

The experiments are designed to be practical investigations that provide the student with a permanent record of data, results, and answers to questions. Each experiment has the following sections:

- **Reading:** A chapter reference to the *Digital Fundamentals: A Systems Approach* text appears in the heading for each experiment.
- **Objectives:** These are statements of what the student should be able to do after completing the experiment.
- **Materials Needed:** This is a list of the components and small items required. Most experiments use TTL logic because it is readily available, inexpensive, and works well in basic protoboards.
- **Summary of Theory:** The Summary of Theory is intended to reinforce the important concepts in the text with a review of the main points prior to the laboratory experience. In most cases, specific practical information needed in the experiment is presented.
- **Procedure:** This section contains a relatively structured set of steps for performing the experiment. Laboratory techniques are given in detail.

- **For Further Investigation:** This section contains specific suggestions for additional related laboratory work with a less rigorous procedure. It can be used as an enhancement to the experiment.

Most experiments include a removable *Report* section with necessary data tables, schematics, and figures. All of the *Report* sections include

- **Conclusion:** The student summarizes the key findings from the experiment.
- **Further Investigation Results:** The student summarizes results from the For Further Investigation in his or her own words.
- **Evaluation and Review Questions:** This section contains questions that require the student to draw conclusions from the laboratory work and check his or her understanding of the concepts. Troubleshooting questions are frequently presented.

Four experiments do not have a Report section and are designed for the student to do a more formal lab report of his or her findings. These experiments are identified with a pencil icon.

Experiments 25 and 26 introduce students to PLD programming using VHDL, Verilog and Schematic capture based software. Compilers are available from different vendors, including Altera (Quartus II), Xilinx (Project Navigator), and ModelSim. The various programming software used in these experiments is the free web edition that can be downloaded from the manufacturer's website. Introductory PLD development tutorials are provided at www.pearsonhighered.com/floyd covering VHDL,

Verilog, and schematic capture techniques. Project files for the experiments are also available at this site.

Experiment 28, entitled *Introduction to the Intel Processors*, is set up as a tutorial on the structure of Intel processors and includes instructions for inputting and running a simple assembly language program using the Debug program. The experiment requires a PC to explore the architecture of microprocessors and provides a "bridge" to a microprocessor course. Spaces are left within the procedure section of this experiment for students' answers and observations.

Most experiments have a set of PowerPoint® slides that are designed as a class review of the experiment and many have troubleshooting problems posed. PowerPoint® slides are available to instructors at http://pearsonhighered.com/floyd/

To access supplementary materials online, instructors need to request an instructor access code. Go to **www.pearsonhighered.com/irc**, where you can register for an instructor access code. Within 48 hours after registering, you will receive a confirming e-mail, including an instructor access code. Once you have received your code, go to the site and log on for full instructions on downloading the materials you wish to use.

Multisim® computer simulation files for 7 experiments in the lab manual are also available on the website for students to download (http://pearsonhighered.com/floyd/). Each of the 7 Multisim circuits operates correctly. A troubleshooting worksheet associated with each circuit is available on the website. The troubleshooting problems are in two forms: 1) A symptom is given; the student deduces the probable fault, and then tests his or her deduction. 2) A fault is described; the student deduces its probable effect and tests his or her deduction. Of course, no computer simulation can replace the practical benefit of actual laboratory work, but a computer simulation has the capability for students to do *what-if* analysis for the lab circuits including simulating effects that might damage an actual circuit. A tutorial is on the website to provide an introduction to Multisim.

Another feature of this manual is a discussion of formal report writing and a tutorial on analog and digital oscilloscopes. Because of the importance of understanding data sheets, abridged data sheets for the various ICs used in the experiments are included in Appendix A. Unabridged data sheets are available from the manufacturer's websites. Appendix B has a complete list of materials for the experiments.

We would like to thank Dan Trudden and Rex Davidson at Pearson Education and Lois Porter, who copyedited the manuscript and made many excellent suggestions along the way. Finally, we express our mutual appreciation for the support of our wives.

David Buchla
Doug Joksch

Introduction to the Student

Circuit Wiring

An important skill needed by electronics technicians is that of transforming circuit drawings into working prototypes. The circuits in this manual can be constructed on solderless protoboards ("breadboards") available at Radio Shack and other suppliers of electronic equipment. These boards use #22- or #24-gauge solid core wire, which should have 3/8 inch of the insulation stripped from the ends. Protoboard wiring is not difficult, but it is easy to make a wiring error that is time-consuming to correct. Wires should be kept neat and close to the board. Avoid wiring across the top of integrated circuits (ICs) or using wires much longer than necessary. A circuit that looks like a plate of spaghetti is difficult to follow and worse to troubleshoot.

One useful technique to help avoid errors, especially with larger circuits, is to make a wire list. After assigning pin numbers to the ICs, tabulate each wire in the circuit, showing where it is to be connected and leaving a place to check off when it has been installed. Another method is to cross out each wire on the schematic as it is added to the circuit. Remember the power supply and ground connections, because they frequently are left off logic drawings. Finally, it is useful to "daisy-chain," in the same color, signal wires that are connected to the same electrical point. Daisy-chaining is illustrated in Figure I–1.

Troubleshooting

When the wiring is completed, test the circuit. If it does not work, turn off the power and recheck the wiring. Wiring, rather than a faulty component, is the more likely cause of an error. Check that the proper power and ground are connected to each IC. If the problem is electrical noise, decoupling capacitors between the power supply and ground may help. Good troubleshooting requires the technician to understand clearly the purpose of the circuit and its normal operation. It can begin at the input and proceed toward the output; or it can begin at the output and proceed toward the input; or it can be done by half-splitting the circuit. Whatever procedure you choose, there is no substitute for understanding how the circuit is supposed to behave and applying your knowledge to the observed conditions in a systematic way.

The Laboratory Notebook

Purpose of a Laboratory Notebook

The laboratory notebook forms a chronologic record of laboratory work in such a manner that it can be reconstructed if necessary. The notebook is a bound and numbered daily record of laboratory work. Data are recorded as they are observed. Each page is dated as it is done and the signature of the person doing the work is added to make the work official; laboratory notebooks may be the basis of

FIGURE I–1

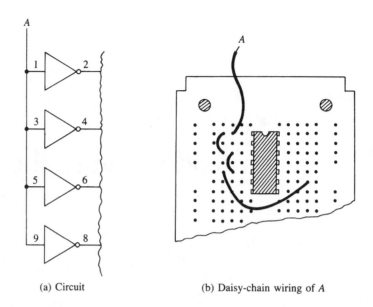

(a) Circuit   (b) Daisy-chain wiring of A

patent applications or have other legal purposes. No pages are left blank and no pages may be removed.

General Information

The format of laboratory notebooks may vary; however, certain requirements are basic to all laboratory notebooks. The name of the experimenter, date, and purpose of the experiment are entered at the top of each page. All test equipment should be identified by function, manufacturer, and serial number to facilitate reconstruction of the experiment. Test equipment may be identified on a block diagram or circuit drawing rather than an equipment list. References to any books, articles, or other sources that were used in preparing for the experiment are noted. A *brief* description of the procedure is necessary. The procedure is not a restatement of the instructions in the experiment book but rather is a concise statement about what *was* done in the experiment.

Recording of Data

Data taken in an experiment should be directly recorded in tabular form in the notebook. Raw (not processed) data should be recorded. They should not be transcribed from scratch paper. When calculations have been applied to data, a sample calculation should be included to indicate clearly what process has been applied to the raw data. If an error is made, a single line should be drawn through the error with a short explanation.

Graphs

A graph is a visual tool that can quickly convey to the reader the relationship between variables. The eye can discern trends in magnitude or slope more easily from graphs than from tabular data. Graphs are constructed with the dependent variable plotted along the horizontal axis (called the *abscissa*) and the independent variable plotted along the vertical axis (called the *ordinate*). A smooth curve can be drawn showing the trend of the data. It is not necessary to connect the data points (except in calibration curves). For data in which one of the variables is related to the other by a power, logarithmic (log) scales in one or both axes may show the relationship of data. Log scales can show data over a large range of values that will not fit on ordinary graph paper.

When you have determined the type of scale that best shows the data, select numbers for the scale that are easily read. Do not use the data for the scale; rather, choose numbers that allow the largest data point to fit on the graph. Scales should generally start from zero unless limitations in the data preclude it. Data points on the graph should be shown with a dot in the center of a symbol such as a circle or triangle. The graph should be labeled in a self-explanatory manner. A figure number should be used as a reference in the written explanation.

Schematics and Block Diagrams

Schematics, block diagrams, waveform drawings, and other illustrations are important tools to depict the facts of an experiment. Experiments with circuits need at least a schematic drawing showing the setup and may benefit from other illustrations depending on your purpose. Usually, simple drawings are best; however, sufficient detail must be shown to enable the reader to reconstruct the circuit if necessary. Adequate information should be contained with an illustration to make its purpose clear to the reader.

Results and Conclusion

The section that discusses the results and conclusion is the most important part of your laboratory notebook; it contains the key findings of your experiment. Each experiment should contain a conclusion, which is a *specific* statement about the important results you obtained. Be careful about sweeping generalizations that are not warranted by the experiment. Before writing a conclusion, it is useful to review the purpose of the experiment. A good conclusion "answers" the purpose of the experiment. For example, if the purpose of the experiment was to determine the frequency response of a filter, the conclusion should describe the frequency response or contain a reference to an illustration of the response. In addition, the conclusion should contain an explanation of difficulties, unusual results, revisions, or any suggestions you may have for improving the circuit.

Suggested Format

From the foregoing discussion, the following format is suggested. This format may be modified as circumstances dictate.

1. *Title and date.*
2. *Purpose:* Give a statement of what you intend to determine as a result of the investigation.
3. *Equipment and materials:* Include equipment model and serial numbers, which can allow retracing if a defective or uncalibrated piece of equipment was used.
4. *Procedure:* Include a description of what you did and what measurements you made. A reference to the schematic drawing should be included.
5. *Data:* Tabulate raw (unprocessed) data; data may be presented in graph form.
6. *Sample calculations:* If you have a number of calculations, give a sample calculation that shows the formulas you applied to the raw data to transform it to processed data. This section may be omitted if calculations are clear from the procedure or are discussed in the results.
7. *Results and conclusion:* This section is the place to discuss your results, including experimental errors. This section should contain key information about the results and your interpretation of the significance of these results.

Each page of the laboratory notebook should be signed and dated, as previously discussed.

The Technical Report

Effective Writing

The purpose of technical reports is to communicate technical information in a way that is easy for the reader to understand. Effective writing requires that you know your reader's background. You must be able to put yourself in the reader's place and anticipate what information you must convey to have the reader understand what you are trying to say. When you are writing experimental results for a person working in your field, such as an engineer, your writing style may contain words or ideas that are unfamiliar to a layperson. If your report is intended for persons outside your field, you will need to provide background information.

Words and Sentences

You will need to either choose words that have clear meaning to a general audience or define every term that does not have a well-established meaning. Keep sentences short and to the point. Short sentences are easiest for the reader to comprehend. Avoid stringing together a series of adjectives or modifiers. For example, the figure caption below contains a jibberish string of modifiers:

Operational amplifier constant current
source schematic

By changing the order and adding natural connectors such as *of, using,* and *an,* the meaning can be clarified:

Schematic of a constant current source using
an operational amplifier

Paragraphs

Paragraphs must contain a unit of thought. Excessively long paragraphs suffer from the same weakness that afflicts overly long sentences. The reader is asked to digest too much material at once, causing comprehension to diminish. Paragraphs should organize your thoughts in a logical format. Look for natural breaks in your ideas. Each paragraph should have one central idea and contribute to the development of the entire report.

Good organization is the key to a well-written report. Outlining in advance will help organize your ideas. The use of headings and subheadings for paragraphs or sections can help steer the reader through the report. Subheadings also prepare the reader for what is ahead and make the report easier to understand.

Figures and Tables

Figures and tables are effective ways to present information. Figures should be kept simple and to the point. Often a graph can make clear the relationship between data. Comparisons of different data drawn on the same graph make the results more obvious to the reader. Figures should be labeled with a figure number and a brief title. Don't forget to label both axes of graphs.

Data tables are useful for presenting data. Usually, data that are presented in a graph or figure should not also be included in a data table. Data tables should be labeled with a table number and short title. The data table should contain enough information to make its meaning clear: The reader should not have to refer to the text. If the purpose of the table is to compare information, then form the data in columns rather than rows. Column information is easier for people to compare. Table footnotes are a useful method of clarifying some point about the data. Footnotes should appear at the bottom of the table with a key to where the footnote applies.

Data should appear throughout your report in consistent units of measurement. Most sciences use the metric system; however, the English system is still sometimes used. The metric system uses derived units, which are cgs (centimeter-gram-second) or mks (meter-kilogram-second). It is best to use consistent metric units throughout your report or to include a conversion chart.

Reporting numbers using powers of 10 can be a sticky point with reference to tables. Table I–1 shows four methods of abbreviating numbers in tabular form. The first column is unambiguous as the number is presented in conventional form. This requires more space than if we present the information in scientific notation. In column 2, the same data are shown with a metric prefix used for the unit. In column 3, the power of 10 is shown. Each of the first three columns shows the measurement unit and is not subject to misinterpretation. Column 4, on the other hand, is wrong. In this case the author is trying to tell us what operation was performed on the numbers to obtain the values in the column. This is incorrect because the column heading should contain the unit of measurement for the numbers in the column.

Suggested Format

1. *Title:* A good title must convey the substance of your report by using key words that provide the reader with enough information to decide whether the report should be investigated further.
2. *Contents:* Key headings throughout the report are listed with page numbers.
3. *Abstract:* The abstract is a *brief* summary of the work, with principal facts and results stated in concentrated form. It is a key factor for a reader to determine if he or she should read further.
4. *Introduction:* The introduction orients your reader to your report. It should briefly state what you did and give the reader a sense of the purpose of the report. It may tell the reader what to expect and briefly describe the report's organization.
5. *Body of the report:* The report can be made clearer to the reader if you use headings and subheadings in your report. The headings and subheadings can be generated from the outline of your report. Figures and tables should be labeled and referenced from the body of the report.
6. *Conclusion:* The conclusion summarizes important points or results. It may refer to figures or tables previously discussed in the body of the report to add emphasis to significant points. In some cases, the primary reasons for the report are contained within the body and a conclusion is deemed to be unnecessary.
7. *References:* References are cited to enable the reader to find information used in developing your report or work that supports your report. The reference should include all authors' names in the order shown in the original document. Use quotation marks around portions of a complete document such as a journal article or a chapter of a book. Books, journals, or other complete documents should be underlined. Finally, list the publisher, city, date, and page numbers.

TABLE I–1

Reporting numbers in tabular data.

Column 1 Resistance (Ω)	Column 2 Resistance (kΩ)	Column 3 Resistance ($\times 10^3\ \Omega$)	Column 4 Resistance ($\Omega \times 10^{-3}$)
470,000	470	470	470
8,200	8.2	8.2	8.2
1,200,000	1,200	1,200	1,200
330	0.33	0.33	0.33

Correct ———————————— Wrong

Oscilloscope Guide
Analog and Digital Storage Oscilloscopes

The oscilloscope is the most widely used general-purpose measuring instrument because it presents a graph of the voltage as a function of time in a circuit. Many circuits have specific timing requirements or phase relationships that can be readily measured with a two-channel oscilloscope. The voltage to be measured is converted into a visible display that is presented on a screen.

There are two basic types of oscilloscope: analog and digital. In general, they each have specific characteristics. Analog scopes are the classic "real-time" instruments that show the waveform on a cathode-ray tube (CRT). Digital oscilloscopes are rapidly replacing analog scopes because of their ability to store waveforms and because of measurement automation and many other features such as connections for computers and printers. The storage function is so important that it is usually incorporated in the name, for example, a Digital Storage Oscilloscope (DSO). Some higher-end DSOs can emulate an analog scope in a manner that blurs the distinction between the two types. Tektronix, for example, has a line of scopes called DPOs (Digital Phosphor Oscilloscopes) that can characterize a waveform with intensity gradients like an analog scope and gives the benefits of a digital oscilloscope for measurement automation.

Both types of scopes have similar functions and the basic controls are essentially the same for both types (although certain enhanced features are not). In the descriptions that follow, the analog scope is introduced first to familiarize you with basic controls, then a basic digital storage oscilloscope is described.

Analog Oscilloscopes

Block Diagram

The analog oscilloscope contains four functional blocks, as illustrated in Figure I–2. Shown within these blocks are the most important typical controls found on nearly all oscilloscopes. Each of two input channels is connected to the vertical section, which can be set to attenuate or amplify the input signals to provide the proper voltage level to the vertical deflection plates of the CRT. In a dual-trace oscilloscope (the most common type), an electronic switch rapidly switches between channels to send one or the other to the display section. The trigger section samples the input waveform and sends a synchronizing trigger signal at the proper time to the horizontal section. The trigger occurs at the same relative time, thus superimposing each succeeding trace on the previous trace. This action causes the signal to appear to stop, allowing you to examine the signal. The horizontal section contains the time-base (or sweep) generator, which produces a linear ramp, or "sweep," waveform that controls the rate the beam moves across the screen. The horizontal position of the beam is proportional to the time that elapsed from the start of the sweep, allowing the horizontal axis to be calibrated in units of time. The output of the horizontal section is applied to the

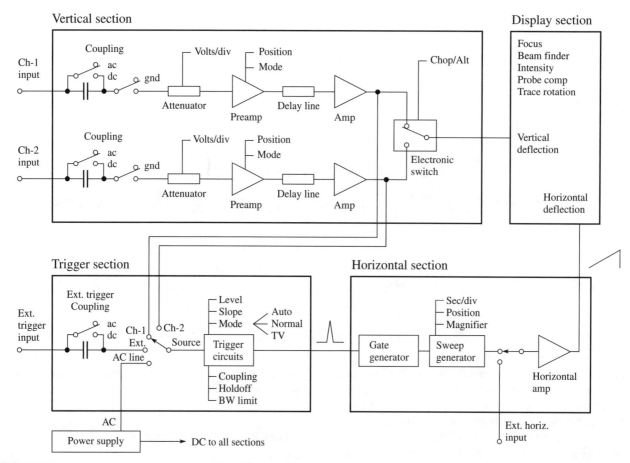

FIGURE I–2
Block diagram of a basic analog oscilloscope.

horizontal deflection plates of the CRT. Finally, the display section contains the CRT and beam controls. It enables the user to obtain a sharp presentation with the proper intensity. The display section usually contains other features such as a probe compensation jack and a beam finder.

Controls

Generally, controls for each section of the oscilloscope are grouped together according to function. Frequently, there are color clues to help you identify groups of controls. Details of these controls are explained in the operator's manual for the oscilloscope; however, a brief description of frequently used controls is given in the following paragraphs. The important controls are shown on the block diagram of Figure I–2.

Display Controls The display section contains controls for adjusting the electron beam, including FOCUS and INTENSITY controls. FOCUS and INTENSITY are adjusted for a comfortable viewing

level with a sharp focus. The display section may also contain the BEAM FINDER, a control used in combination with the horizontal and vertical POSITION controls to bring the trace on the screen. Another control over the beam intensity is the z-axis input. A control voltage on the z-axis input can be used to turn the beam on or off or adjust its brightness. Some oscilloscopes also include the TRACE ROTATION control in the display section. TRACE ROTATION is used to align the sweep with a horizontal graticule line. This control is usually adjusted with a screwdriver to avoid accidental adjustment. Usually a PROBE COMP connection point is included in the display group of controls. Its purpose is to allow a quick qualitative check on the frequency response of the probe-scope system.

Vertical Controls The vertical controls include the VOLTS/DIV (vertical sensitivity) control and its vernier, the input COUPLING switch, and the vertical POSITION control. There is a duplicate set of these controls for each channel and various switches for selecting channels or other vertical operating modes. The vertical input is connected through a selectable attenuator to a high input impedance dc amplifier. The VOLTS/DIV control on each channel selects a combination of attenuation/gain to determine the vertical sensitivity. For example, a low-level signal will need more gain and less attenuation than a higher level signal. The vertical sensitivity is adjusted in fixed VOLTS/DIV increments to allow the user to make calibrated voltage measurements. In addition, a concentric vernier control is usually provided to allow a continuous range of sensitivity. This knob must be in the detent (calibrated) position to make voltage measurements. The detent position can be felt by the user as the knob is turned because the knob tends to "lock" in the detent position. Some oscilloscopes have a warning light or message when the vernier is not in its detent position.

The input coupling switch is a multiple-position switch that can be set for AC, GND, or DC and sometimes includes a 50 Ω position. The GND position of the switch internally disconnects the signal from the scope and grounds the input amplifier. This position is useful if you want to set a ground reference level on the screen for measuring the dc component of a waveform. The AC and DC positions are high-impedance inputs—typically 1 MΩ shunted by 15 pF of capacitance. High-impedance inputs are useful for general probing at frequencies below about 1 MHz. At higher frequencies, the shunt capacitance can load the signal source excessively, causing measurement error. Attenuating divider probes are good for high-frequency probing because they have very high impedance (typically 10 MΩ) with very low shunt capacitance (as low as 2.5 pF).

The AC position of the coupling switch inserts a series capacitor before the input attenuator, causing dc components of the signal to be blocked. This position is useful if you want to measure a small ac signal riding on top of a large dc signal-power supply ripple, for example. The DC position is used when you want to view both the AC and DC components of a signal. This position is best when viewing digital signals, because the input *RC* circuit forms a differentiating network. The AC position can distort the digital waveform because of this differentiating circuit. The 50 Ω position places an accurate 50 Ω load to ground. This position provides the proper termination for probing in 50 Ω systems and reduces the effect of a variable load, which can occur in high-impedance termination. The effect of source loading must be taken into account when using a 50 Ω input. It is important not to overload the 50 Ω input, because the resistor is normally rated for only 2 W—implying a maximum of 10 V of signal can be applied to the input.

The vertical POSITION control varies the dc voltage on the vertical deflection plates, allowing you to position the trace anywhere on the screen. Each channel has its own vertical POSITION control, enabling you to separate the two channels on the screen. You can use vertical POSITION when the coupling switch is in the GND position to set an arbitrary level on the screen as ground reference.

There are two types of dual-channel oscilloscope: dual beam and dual trace. A dual-beam oscilloscope has two independent beams in the CRT and independent vertical deflection systems, allowing both signals to be viewed at the same time. A dual-trace oscilloscope has only one beam and one deflection system; it uses electronic switching to show the two signals. Dual-beam oscilloscopes are generally restricted to high-performance research instruments and are much more expensive than dual-trace oscilloscopes. The block diagram in Figure I–2 is for a typical dual-trace oscilloscope.

A dual-trace oscilloscope has user controls labeled CHOP or ALTERNATE to switch the beam between the channels so that the signals appear to occur simultaneously. The CHOP mode rapidly switches the beam between the two channels at a fixed high speed rate, so the two channels appear to be displayed at the same time. The ALTERNATE mode first completes the sweep for one of the channels and then displays the other channel on the next (or alternate) sweep. When viewing slow signals, the CHOP mode is best because it reduces the flicker that would otherwise be observed. High-speed signals can usually be observed best in ALTERNATE mode to avoid seeing the chop frequency.

Another feature on most dual-trace oscilloscopes is the ability to show the algebraic sum and difference of the two channels. For most measurements, you should have the vertical sensitivity (VOLTS/DIV) on the same setting for both channels. You can use the algebraic sum if you want to compare the balance on push-pull amplifiers, for example. Each amplifier should have identical out-of-phase signals. When the signals are added, the resulting display should be a straight line, indicating balance. You can use the algebraic difference when you want to measure the waveform across an ungrounded component. The probes are connected across the ungrounded component with probe ground connected to circuit ground. Again, the vertical sensitivity (VOLTS/DIV) setting should be the same for each channel. The display will show the algebraic difference in the two signals. The algebraic difference mode also allows you to cancel any unwanted signal that is equal in amplitude and phase and is common to both channels.

Dual-trace oscilloscopes also have an X-Y mode, which causes one of the channels to be graphed on the X-axis and the other channel to be graphed on the Y-axis. This is necessary if you want to change the oscilloscope baseline to represent a quantity other than time. Applications include viewing a transfer characteristic (output voltage as a function of input voltage), swept frequency measurements, or showing Lissajous figures for phase measurements. Lissajous figures are patterns formed when sinusoidal waves drive both channels.

Horizontal Controls The horizontal controls include the SEC/DIV control and its vernier, the horizontal magnifier, and the horizontal POSITION control. In addition, the horizontal section may include delayed sweep controls. The SEC/DIV control sets the sweep speed, which controls how fast the electron beam is moved across the screen. The control has a number of calibrated positions divided into steps of 1-2-5 multiples, which allow you to set the exact time interval at which you view the input signal. For example, if the graticule has 10 horizontal divisions and the SEC/DIV control is set to 1.0 ms/div, then the screen will show a total time of 10 ms. The SEC/DIV control usually has a concentric vernier control that allows you to adjust the sweep speed continuously between the calibrated steps. This control must be in the detent position in order to make calibrated time measurements. Many scopes are also equipped with a horizontal magnifier that affects the time base. The magnifier increases the sweep time by the magnification factor, giving you increased resolution of signal details. Any portion of the original sweep can be viewed using the horizontal POSITION control in conjunction with the magnifier. This control actually speeds the sweep time by the magnification factor and therefore affects the calibration of the time base set on the SEC/DIV control. For example, if you are using a 10× magnifier, the SEC/DIV dial setting must be divided by 10.

Trigger Controls The trigger section is the source of most difficulties when learning to operate an oscilloscope. These controls determine the proper time for the sweep to begin in order to produce a stable display. The trigger controls include the MODE switch, SOURCE switch, trigger LEVEL, SLOPE, COUPLING, and variable HOLDOFF controls. In addition, the trigger section includes a connector for applying an EXTERNAL trigger to start the sweep. Trigger controls may also include HIGH or LOW FREQUENCY REJECT switches and BANDWIDTH LIMITING.

The MODE switch is a multiple-position switch that selects either AUTO or NORMAL (sometimes called TRIGGERED) and may have other positions, such as TV or SINGLE sweep. In the AUTO position, the trigger generator selects an internal oscillator that will trigger the sweep generator as long as no other trigger is available. This mode ensures that a sweep will occur even in the absence of a signal, because the trigger circuits will "free run" in this mode. This allows you to obtain a baseline for adjusting ground reference level or for adjusting the display controls. In the NORMAL or TRIGGERED mode, a trigger is generated from one of three sources selected by the SOURCE switch: the INTERNAL signal, an EXTERNAL trigger source, or the AC LINE. If you are using the internal signal to obtain a trigger, the normal mode will provide a trigger only if a signal is present and other trigger conditions (level, slope) are met. This mode is more versatile than AUTO as it can provide stable triggering for very low to very high frequency signals. The TV position is used for synchronizing either television fields or lines and SINGLE is used primarily for photographing the display.

The trigger LEVEL and SLOPE controls are used to select a specific point on either the rising or falling edge of the input signal for generating a trigger. The trigger SLOPE control determines which edge will generate a trigger, whereas the LEVEL control allows the user to determine the voltage level on the input signal that will start the sweep circuits.

The SOURCE switch selects the trigger source—either from the CH-1 signal, the CH-2 signal, an EXTERNAL trigger source, or the AC LINE. In the CH-1 position, a sample of the signal from channel-1 is used to start the sweep. In the EXTERNAL position, a time-related external signal is used for triggering. The external trigger can be coupled with either AC or DC COUPLING. The trigger signal can be coupled with AC COUPLING if the trigger signal is riding on a dc voltage. DC COUPLING is used if the triggers occur at a frequency of less than about 20 Hz. The AC LINE position causes the trigger to be derived from the ac power source. This synchronizes the sweep with signals that are related to the power line frequency.

The variable HOLDOFF control allows you to exclude otherwise valid triggers until the holdoff time has elapsed. For some signals, particularly complex waveforms or digital pulse trains, obtaining a stable trigger can be a problem. This can occur when one or more valid trigger points occurs before the signal repetition time. If every event that the trigger circuits qualified as a trigger were allowed to start a sweep, the display could appear to be unsynchronized. By adjusting the variable HOLDOFF control, the trigger point can be made to coincide with the signal-repetition point.

Oscilloscope Probes

Signals should always be coupled into an oscilloscope through a probe. A probe is used to pick off a signal and couple it to the input with a minimum loading effect on the circuit under test. Various types of probes are provided by manufacturers, but the most common type is a 10:1 attenuating probe that is shipped with most general-purpose oscilloscopes. These probes have a short ground lead that should be connected to a nearby circuit ground point to avoid oscillation and power line interference. The ground lead makes a mechanical connection to the test circuit and passes the signal through a flexible, shielded cable to the oscilloscope. The shielding helps protect the signal from external noise pickup.

Begin any session with the oscilloscope by checking the probe compensation on each channel. Adjust the probe for a flat-topped square wave while observing the scope's calibrator output. This is a good signal to check the focus and intensity and verify trace alignment. Check the front-panel controls for the type of measurement you are going to make. Normally, the variable controls (VOLTS/DIV and SEC/DIV) should be in the calibrated (detent) position. The vertical coupling switch is usually placed in the DC position unless the waveform you are interested in has a large dc offset. Trigger holdoff should be in the minimum position unless it is necessary to delay the trigger to obtain a stable sweep.

Digital Storage Oscilloscopes

Block Diagram

The digital storage oscilloscope (DSO) uses a fast analog-to-digital converter (ADC) on each channel (typically two or four channels) to convert the input voltage into numbers that can be stored in a memory. The digitizer samples the input at a uniform rate called the sample rate; the optimum sample rate depends on the speed of the signal. The process of digitizing the waveform has many advantages for accuracy, triggering, viewing hard to see events, and for waveform analysis. Although the method of acquiring and displaying the waveform is quite different than analog scopes, the basic controls on the instrument are similar.

A block diagram of the basic DSO is shown in Figure I–3. As you can see, functionally, the block diagram is like that of the analog scope. As in the analog oscilloscope, the vertical and horizontal controls include position and sensitivity that are used to set up the display for the proper scaling.

Specifications Important parameters with DSOs include the resolution, maximum digitizing rate, and the size of the acquisition memory as well as the available analysis options. The resolution is determined by the number of bits digitized by the ADC. A low resolution DSO may use only six bits (one part in 64). A typical DSO may use 8 bits, with each channel sampled simultaneously. High-end DSOs may use 12 bits. The maximum digitizing rate is important to capture rapidly changing signals; typically the maximum rate is 1 GSample/s. The size of the memory determines the length of time the sample can be taken; it is also important in certain waveform measurement functions.

Triggering One useful feature of digital storage oscilloscopes is their ability to capture waveforms either before or after the trigger event. Any segment of the waveform, either before or after the trigger event, can be captured for analysis. **Pretrigger capture** refers to acquisition of data that occurs *before* a trigger event. This is possible because the data is digitized continuously, and a trigger event can be selected to stop the data collection at some point in the sample window. With pretrigger capture, the scope can be triggered on the fault condition, and the signals that preceded the fault condition can be observed. For example, troubleshooting an occasional glitch in a system is one of the most difficult troubleshooting jobs; by employing pretrigger capture, trouble leading to the fault can be analyzed. A similar application of pretrigger capture is in material failure studies where the events leading to failure are most interesting, but the failure itself causes the scope triggering.

Besides pretrigger capture, posttriggering can also be set to capture data that occurs some time after a trigger event. The record that is acquired can begin after the trigger event by some amount of time or by a specific number of events as determined by a counter. A low-level response to a strong stimulus

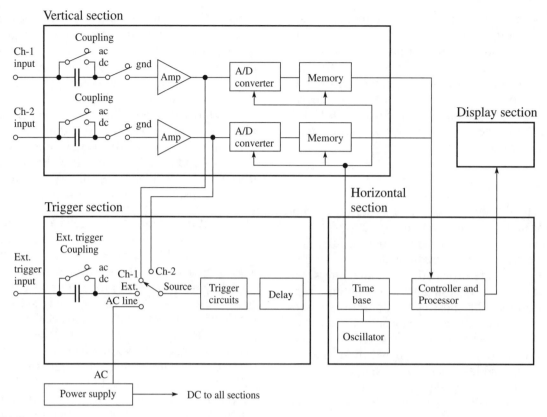

FIGURE I–3
Block diagram of a basic digital storage oscilloscope.

signal is an example of when posttriggering is useful.

Because of the large number of functions that can be accomplished by even basic DSOs, manufacturers have largely replaced the plethora of controls with menu options, similar to computer menus and detailed displays that show the controls as well as measurement parameters. CRTs have been replaced by liquid crystal displays, similar to those on laptop computers. As an example, the display for a basic digital storage oscilloscope is shown in Figure I–4. Although this is a basic scope, the information available to the user right at the display is impressive.

The numbers on the display in Figure I–4 refer to the following parameters:

1. Icon display shows acquisition mode.

        Sample mode

        Peak detect mode

        Average mode

2. Trigger status shows if there is an adequate trigger source or if the acquisition is stopped.
3. Marker shows horizontal trigger position. This also indicates the horizontal position since the Horizontal Position control actually moves the trigger position horizontally.
4. Trigger position display shows the difference (in time) between the center graticule and the trigger position. Center screen equals zero.
5. Marker shows trigger level.
6. Readout shows numeric value of the trigger level.
7. Icon shows selected trigger slope for edge triggering.
8. Readout shows trigger source used for triggering.
9. Readout shows window zone timebase setting.
10. Readout shows main timebase setting.
11. Readout shows channels 1 and 2 vertical scale factors.
12. Display area shows on-line messages momentarily.
13. On-screen markers show the ground reference points of the displayed waveforms. No marker indicates the channel is not displayed.

FIGURE I–4
The display area for a basic digital oscilloscope.

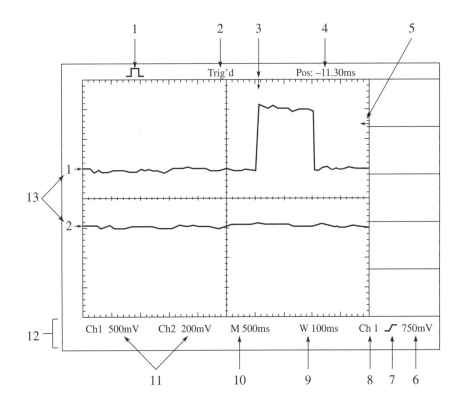

A front view of a Tektronix TDS2024 is shown in Figure I–5. Operation is similar to that of an analog scope except more of the functions are menu controlled. For example, the MEASURE function brings up a menu that the user can select from automated measurements including voltage, frequency, period, and averaging.

FIGURE I–5
The Tektronix TDS2024 oscilloscope (photo courtesy of Tektronix, Inc.).

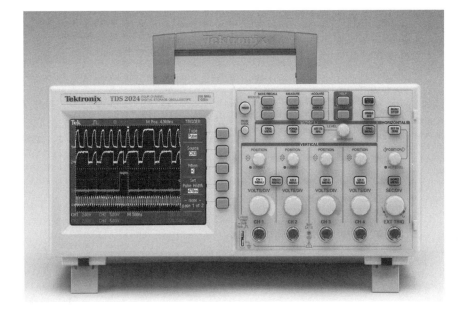

Experiment 1
Laboratory Instrument Familiarization

Objectives

After completing this experiment, you will be able to

☐ Use a digital multimeter (DMM) to measure a specified dc voltage from the power supply.

☐ Use an oscilloscope to measure circuit voltages and frequencies.

☐ Set up the function generator to obtain a transistor-transistor logic (TTL) compatible pulse of a specified frequency. Measure the pulse amplitude and the frequency with an oscilloscope.

☐ Construct a digital oscillator circuit on a laboratory protoboard and measure various parameters with the oscilloscope.

Materials Needed

Light-emitting diode (LED)
Resistors: one 330 Ω, one 1.0 kΩ, one 2.7 kΩ
Capacitors: one 0.1 μF, one 100 μF
One 555 timer

For Further Investigation:
  Current tracer
  Logic pulser
  One 100 Ω resistor

Summary of Theory

Laboratory equipment needed for most electronics work includes a DMM, a power supply, a function generator, and a dual-trace analog or digital oscilloscope. This experiment is an introduction to these instruments and to protoboards that are commonly used to wire laboratory experiments. Since each laboratory will have instruments from different manufacturers and different models, you should familiarize yourself with your particular lab station using the manufacturer's operating instructions or information provided by your instructor. There is a wide variety of instruments used in electronics labs; however, the directions in this experiment are general enough that you should be able to follow them for whatever instruments you are using.

The Power Supply

All active electronic devices, such as the integrated circuits used in digital electronics, require a stable source of dc voltage to function properly. The power supply provides the proper level of dc voltage. It is very important that the correct voltage be set before connecting it to the ICs on your board or permanent damage can result. The power supply at your bench may have more than one output and normally will have a built-in meter to help you set the voltage. For nearly all of the circuits in this manual, the power supply should be set to +5.0 V. When testing a faulty circuit, one of the first checks is to verify that the supply voltage is correct and that there is no ac component to the power supply output.

The Digital Multimeter

The DMM is a multipurpose measuring instrument that combines in one instrument the characteristics of a dc and ac voltmeter, a dc and ac ammeter, and an ohmmeter. The DMM indicates the measured quantity as a digital number, avoiding the necessity to interpret the scales as was necessary on older instruments.

Because the DMM is a multipurpose instrument, it is necessary to determine which controls select the desired function. In addition, current measurements (and often high-range voltage measurements) usually require a separate set of lead connections to the meter. After you have selected the function, you may need to select the appropriate range to make the measurement. Digital multimeters can be autoranging, meaning that the instrument automatically selects the correct scale and sets the decimal place, or they can be manual ranging, meaning that the user must select the correct scale.

The voltmeter function of a DMM can measure either ac or dc volts. For digital work, the dc volts function is always used to verify that the dc supply voltage is correct and to check steady-state logic levels. If you are checking a power supply, you can verify that there is no ac component in the supply voltage by selecting the ac function. With ac voltage selected, the reading of a power supply should be very close to zero. Except for a test like this, the ac voltage function is not used in digital work.

The ohmmeter function of a DMM is used only in circuits that are not powered. When measuring resistance, the power supply should be disconnected from the circuit to avoid measuring the resistance of the power supply. An ohmmeter works by inserting a small test voltage into a circuit and measuring the resulting current flow. Consequently, if any voltage is present, the reading will be in error. The meter will show the resistance of all possible paths between the probes. If you want to know the resistance of a single component, you must isolate that component from the remainder of the circuit by disconnecting one end. In addition, body resistance can affect the reading if you are holding the con-

ducting portion of both probes in your fingers. This procedure should be avoided, particularly with high resistances.

The Function Generator

The function generator is used to produce signals required for testing various kinds of circuits. For digital circuits, a periodic rectangular pulse is the basic signal used for testing logic circuits. It is important that the proper voltage level be set up before connecting the function generator to the circuit or else damage may occur. Function generators normally have controls for adjusting the peak amplitude of a signal and may also have a means of adjusting the 0 volt level. Most function generators have a separate pulse output for use in logic circuits. If you have a TTL compatible output, it will be the one used for the experiments in this manual.

A periodic rectangular pulse is a signal that rises from one level to another level, remains at the second level for a time called the pulse width, (t_w), and then returns to the original level. Important parameters for these pulses are illustrated in Figure 1–1. For digital testing, it is useful to use a duty cycle that is *not* near 50% so that an inverted signal can be readily detected on an oscilloscope.

In addition to amplitude and dc offset controls, function generators have switches that select the range of the output frequency. A vernier control may be present for fine frequency adjustments.

The Oscilloscope

The oscilloscope is the most important test instrument for testing circuits, and you should become completely familiar with its operation. It is a versatile test instrument, letting you "see" a graph of the voltage as a function of time in a circuit and compare waves. Because an oscilloscope allows you to measure various parameters, it is considered to be an instrument capable of parametric measurements important in both digital and analog work. Nearly

FIGURE 1–1
Definitions for a periodic pulse train.

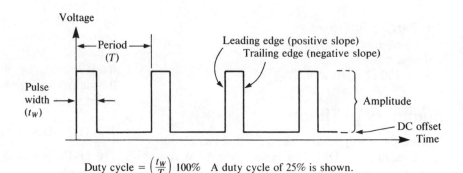

Duty cycle = $\left(\frac{t_W}{T}\right)$ 100%   A duty cycle of 25% is shown.

all complex digital circuits have specific timing requirements that can be readily measured with a two-channel oscilloscope.

There are two basic types of oscilloscopes: analog and digital. Because of its versatility, accuracy, and ability to do automated measurements, digital scopes are the choice of many technicians today. Both types of scopes have four main control groups: display controls, vertical and horizontal controls, and trigger controls. If you are not familiar with these controls, or the operation of the oscilloscope in general, you should read the Oscilloscope Guide starting on page xiii. Both analog and digital scopes are covered in this summary. In addition, you may want to review the operator's manual that came with the oscilloscope at your lab station.

Logic Pulser and Current Tracer

The logic pulser and current tracer are simple digital test instruments that are useful for finding certain difficult faults, such as a short between V_{CC} and ground. A problem like this can be very difficult to find in a large circuit because the short could be located in many possible places. The current tracer responds to pulsing current by detecting the changing magnetic field. A handheld logic pulser can provide very short duration, nondestructive pulses into the shorted circuit. The current tracer, used in conjunction with the pulser or other pulsating source, allows you to follow the current path, leading you directly to the short. This method of troubleshooting is also useful for "stuck" nodes in a circuit (points that have more than one path for current). The sensitivity of the current tracer can be varied over a large range to allow you to trace various types of faults.

Logic Probe

Another handheld instrument that is useful for tracing simple logic circuits is the logic probe. The logic probe can be used to determine the logic level of a point in a circuit or to determine whether there is pulse activity at the point by LED (light-emitting diode) displays. Although it is used primarily for simple circuits because it cannot show important time relationships between digital signals, a good probe can indicate activity on the line, even if it is short pulses. A simple logic probe can determine if logic levels are HIGH, LOW, or INVALID.

Logic Analyzer

One of the most powerful and widely used instruments for digital troubleshooting is the logic analyzer. The logic analyzer is an instrument that originally was designed to make functional (as opposed to parametric) measurements in logic circuits. It is useful for observing the time relationship between a number of digital signals at the same time, allowing the technician to see a variety of errors, including "stuck" nodes, very short noise spikes, intermittent problems, and timing errors. Newer analyzers can include multiple channels of a digital storage oscilloscope (DSO) as well as logic channels. An example of a two-function analyzer that can be equipped with multiple channels of DSO and as many as 680 logic analyzer channels is the Tektronix TLA700 series shown in Figure 1–2. Not all electronic laboratories are equipped with a logic analyzer, even a simple one, and one is not necessary for the experiments in this manual. Further information on logic analyzers is given on various websites on the Internet (see www.Tektronix.com for example).

Protoboards

Protoboards are a convenient way to construct circuits for testing and experimenting. While there are some variations in the arrangement of the hole patterns, most protoboards are similar to the one shown in Figure 1–3, which is modeled after the Radio Shack board 276-174. Notice that the top and bottom horizontal rows are connected as a continuous row.

FIGURE 1–2
Tektronix logic analyzers (courtesy of Tektronix, Inc.).

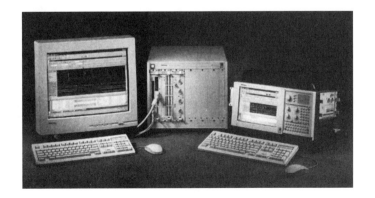

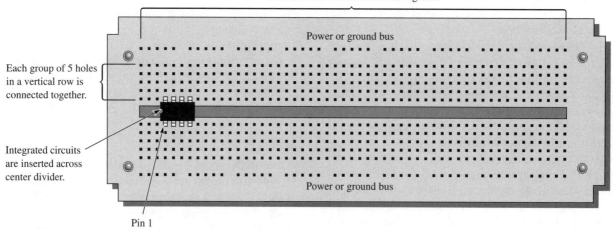

This horizontal row is connected together.

Power or ground bus

Each group of 5 holes in a vertical row is connected together.

Integrated circuits are inserted across center divider.

Power or ground bus

Pin 1

FIGURE 1–3

Protoboard. An 8-pin integrated circuit is shown inserted into the board.

Vertical groups of five holes are connected together; the vertical group above the center strip is not connected to the vertical group below the center strip. The holes are 0.1 inch apart, which is the same spacing as the pins on an integrated circuit DIP (dual in-line pins). Integrated circuits (ICs) are inserted to straddle the center; in this manner, wires can be connected to the pins of the IC by connecting them to the same vertical group as the desired pin.

Pin Numbering

Integrated circuits come in various "packages" as explained in the text. In this manual, you will be using all "DIP chips". To determine the pin numbers, you need to locate pin 1 by looking for a notch or dot on one end (see Figure 1–3). Pin 1 is adjacent to this notch as shown. The numbering for a DIP chip always is counterclockwise from pin 1.

Prototyping System

In many engineering and educational laboratories, the instruments described previously are combined into a data acquisition system that can collect and measure signals and show the results on a computer display. Systems like this are a complete prototyping system integrated into a workstation. An example is the National Instruments ELVIS system and data acquisition device, shown in Figure 1–4. The workstation has all of the instruments built in and a modular protoboard mounted on top.

FIGURE 1–4

Prototyping system (courtesy of National Instruments).

1. Laptop Computer
2. USB Cable
3. NI USB M Series with Mass Termination Device
4. NI USB M Series Device Power Cord
5. Shielded Cable to M Series Device
6. NI ELVIS Benchtop Workstation

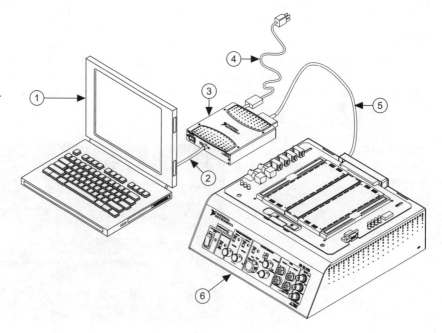

Procedure

Measurement of DC Voltage with the DMM

1. Review the operator's manual or information supplied by your instructor for the power supply at your lab station. Generally, power supplies have a meter or meters that enable you to set the output voltage and monitor the current. Set the voltage based on the power supply meter to +5.0 V and record the reading in Table 1–1 (in the Report section).

2. The +5.0 V is the voltage you will use for nearly all of the experiments in this manual. For most TTL circuits, the power supply should be from 4.75 V to 5.25 V. To check that you have correctly set up the supply, measure the voltage with the DMM. Record the reading of the DMM in Table 1–1.

Measurement of DC Voltage with the Oscilloscope

3. In this step, you will confirm the dc voltage from the power supply using the oscilloscope. Set the SEC/DIV control of your oscilloscope to a convenient value (a value near 0.2 ms/div is suggested to give a steady line on the display). Set the trigger controls to AUTO and INT (internal trigger) to assure a sweep is on the display. Select channel 1 as the input channel, and connect a scope probe to the vertical input. Put the input coupling control on GND to disconnect the input signal and find the ground position on the oscilloscope (digital scopes may have a marker for the GND level, as illustrated in item 13 of Figure I–4). Adjust the beam for a sharp, horizontal line across the scope face.

4. Since you will be measuring a positive voltage, position the ground on a convenient graticule line near the bottom of the display using the vertical POSITION control. If you are using an analog scope, check that the vertical VOLTS/DIV variable knobs are in their calibrated positions. A digital scope, such as the Tektronix TDS2024C, is always calibrated, and there is no vernier control.

5. Move the channel 1 input coupling control from the GND position to the dc position. For almost all digital work, the input coupling control should be in the DC position. Clip the ground lead of the scope probe to the ground of the power supply and touch the probe itself to the power supply output. The line on the face of the oscilloscope should jump up 5 divisions. You can determine the dc voltage by multiplying the vertical sensitivity (1.0 V/div) by the number of divisions observed between ground and this line (5 divisions). Record the measured voltage (to the nearest 0.1 V) in Table 1–1.

Measurement of Pulses with the Oscilloscope

6. Now you will set up the function generator or pulse generator for a logic pulse and measure some characteristics of the pulse using the oscilloscope. Review the operator's manual or information supplied by your instructor for the function generator at your lab station. Select the pulse function and set the frequency for 1.0 kHz. (If you do not have a pulse function, a square wave may be substituted.)

7. Set up and measure the pulse amplitude of the function generator. The vertical sensitivity (VOLTS/DIV) control of the oscilloscope should be set for 1.0 V/div and the SEC/DIV should be left at 0.2 ms/div. Check that both controls are in their calibrated positions. Check the ground level on the oscilloscope as you did in Step 3 and set it for a convenient graticule near the bottom of the scope face. Switch the scope back to dc coupling and clip the ground lead of the scope probe to a ground on the generator. Touch the probe to the function generator's pulse output. If the generator has a variable amplitude control, adjust it for a 4.0 V pulse (4 divisions of deflection). Some generators have a separate control to adjust the dc level of the pulse; others do not. If your generator has a dc offset control, adjust the ground level of the pulse for zero volts.

8. You should obtain a stable display that allows you to measure both the time information and the voltage parameters of the waveform. (If the waveform is not stable, check triggering controls.) In Plot 1 of your report, sketch the observed waveform on the scope display. It is a good idea, whenever you sketch a waveform from a scope, to record the VOLTS/DIV and SEC/DIV settings of controls next to the sketch and to show the ground level. Measure the pulse width (t_w), period (T), and amplitude of the waveform and record these values in Table 1–2. The amplitude is defined in Figure 1–1 and is measured in volts.

9. Connect the LED and series-limiting resistor, R_1, to the pulse generator as shown in Figure 1–5. Note that the LED is a polarized component and must be connected in the correct direction to work. The schematic and an example of protoboard wiring are shown. Measure the signal across the LED with the oscilloscope and show it in Plot 2 of your report. Label the scope settings as in step 8 and show the ground level.

10. Sometimes it is useful to use an oscilloscope to measure the voltage across an ungrounded component. The current-limiting resistor, R_1, in Figure 1–5 is an ungrounded component. To measure the voltage across it, connect both channels of your oscilloscope as shown in Figure 1–6. Make sure that both channels are calibrated and that the

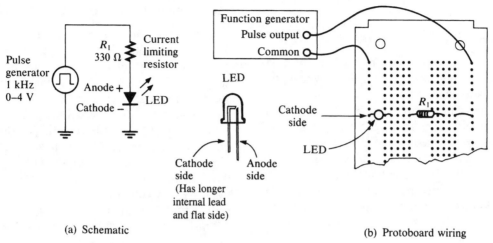

(a) Schematic

(b) Protoboard wiring

FIGURE 1–5
Circuit for Step 9.

vertical sensitivity (VOLTS/DIV) is 1 V/div for each channel. If you are using a newer scope, the difference operation (Channel 1—Channel 2) is likely to be shown as a menu item. On older scopes, the difference measurement is done by inverting channel 2 and selecting the ADD function. Consult the operator's manual if you are not sure. Measure the signal across R_1 and show the result on Plot 3. As a check, the sum of the voltages across the LED and resistor should be equal to the voltage of the generator.

Constructing and Measuring Parameters in a Digital Circuit

11. In this step, you will construct a small digital oscillator. This oscillator generates pulses that could be used to drive other digital circuits. The basic integrated circuit for the oscillator is the 555 timer, which will be covered in detail later. The schematic and sample protoboard wiring is shown in Figure 1–7. Construct the circuit as shown.

12. Using your oscilloscope, observe the signal on pin 3. Sketch the observed signal in Plot 4. Be sure to label the plot with the scope settings (VOLTS/DIV and SEC/DIV). Measure the parameters listed in the first four rows of Table 1–3. The frequency is computed from the period measurement ($f = 1/T$).

13. Replace C_1 with a 100 μF capacitor. The light should blink at a relatively slow rate. A slow frequency like this is useful for visual tests of a circuit or for simulating the opening and closing of a manual switch. Measure the period and frequency

FIGURE 1–6
Measuring an un-grounded component. Both channels must be calibrated and have the same vertical sensitivity settings. On the TDS2024C, the difference between the two channels is on the MATH functions menu.

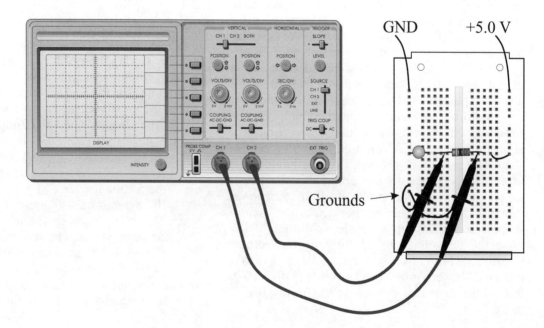

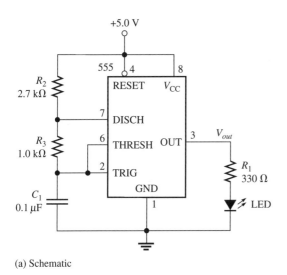

(a) Schematic

(b) Protoboard wiring

FIGURE 1–7
Digital oscillator.

of the oscillator with the 100 μF capacitor. This signal, with a low frequency like this, may give you difficulty if you are using an analog scope. You will need to use NORMAL triggering instead of auto triggering and you may need to adjust the trigger LEVEL control to obtain a stable display. Record your measured values in Table 1–3.

For Further Investigation

Using the Current Tracer

If you have a current tracer available, you can test the paths for current in a circuit such as the one you constructed in step 9. The current tracer can detect the path of pulsing current, which you can follow.

The current tracer detects fast current pulses by sensing the changing magnetic field associated with them. It cannot detect dc. Set the generator to a 1.0 kHz TTL level pulse for this test.

Power the current tracer using a +5.0 V power supply. You will need to provide a common ground for the pulse generator and the power supply, as shown in Figure 1–8(a). (Note that the current tracer has a red wire in one of the leads, which should be connected to the +5.0 V source, and a black wire in the other lead, which should be connected to the common.) The sensitivity of the tracer is adjusted with a variable control near the tip of the current tracer. The current tracer must be held perpendicularly with respect to the conductor in which you are sensing current. In addition, the small mark on the

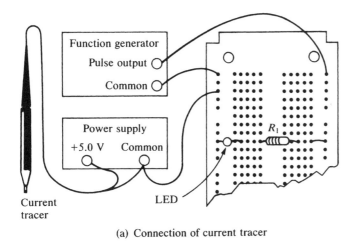

(a) Connection of current tracer

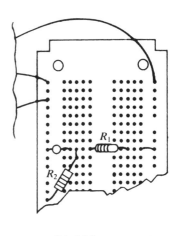

(b) Adding a current path through R_2

FIGURE 1–8

7

FIGURE 1–9

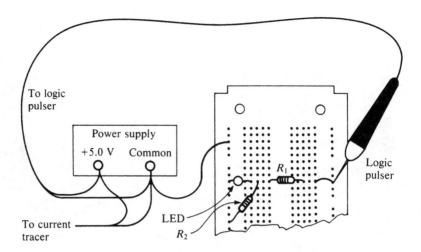

probe tip must be aligned with the current path to obtain maximum sensitivity.

Begin by holding the current tracer above R_1. Rotate the current tracer so that the tip is aligned with the path of current. Adjust the sensitivity to about half-brightness. You should now be able to trace the current path through R_1, the LED, and along the protoboard. Practice tracing the path of current. Simulate a low-impedance fault by installing a 100 Ω resistor (call this R_2 for this experiment) in parallel with the LED, as shown in Figure 1–8(b). Test the circuit with the current tracer to determine the path for current. Does most of the current go through R_2 or through the LED?

Using the Current Tracer and Logic Pulser

Circuit boards typically have many connections where a potential short can occur. If a short occurs between the power supply and ground due to a solder splash or other reason, it can be difficult to find. A logic pulser, used in conjunction with a

current tracer, can locate the fault without the need for applying power to the circuit. The logic pulser applies very fast pulses to a circuit under test. A flashing LED in the tip indicates the output mode, which can be set to various pulse streams or to a continuous series of pulses. The pulser can be used in an operating circuit without damaging it because the energy supplied to the circuit is limited.

Start with the logic pulser by setting it for continuous pulses. Remove the pulse generator from the test circuit and touch the logic pulser to the test circuit, as shown in Figure 1–9. You can hold the current tracer at a 90-degree angle and against the tip of the logic pulser in order to set the sensitivity of the current tracer. You should now be able to follow the path for current as you did before.

Now simulate a direct short fault across the circuit by connecting a wire as shown in Figure 1–10. You may need to adjust the sensitivity of the current tracer. Use the logic pulser and current tracer to follow the path of current. Can you detect current in R_1? Describe in your report the current path in the wire and in the protoboard.

FIGURE 1–10

Simulating a short circuit. The logic pulser forces current through the short; this current can be detected with the current tracer.

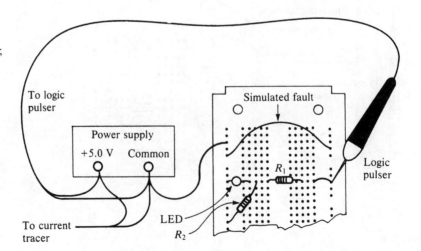

Report for Experiment 1

Name: _____ Date: _____ Class: _____

Objectives:

- ☐ Use a digital multimeter (DMM) to measure a specified dc voltage from the power supply.
- ☐ Use an oscilloscope to measure circuit voltages and frequencies.
- ☐ Set up the function generator to obtain a transistor-transistor logic (TTL) compatible pulse of a specified frequency. Measure the pulse amplitude and the frequency with an oscilloscope.
- ☐ Construct a digital oscillator circuit on a laboratory protoboard and measure various parameters with the oscilloscope.

Data and Observations:

TABLE 1–1

Voltage Setting = 5.0 V	Voltage Reading
Power supply meter	
DMM	
Oscilloscope	

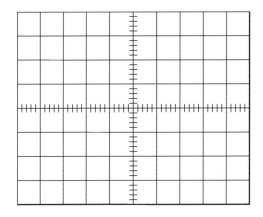

PLOT 1   Generator waveform

TABLE 1–2

Function Generator Parameters (at 1.0 kHz)	Measured Values
Pulse width	
Period	
Amplitude	

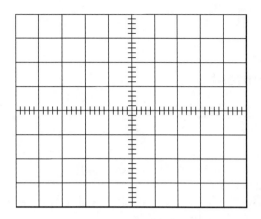

PLOT 2   Voltage across LED

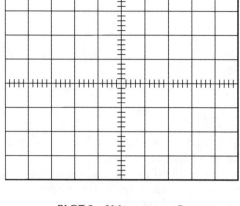

PLOT 3   Voltage across R_1

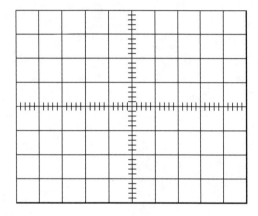

PLOT 4   Digital oscillator output (pin 3)

TABLE 1–3

Step	Digital Oscillator Parameters	Measured Values
12	Period	
	Duty cycle	
	Amplitude	
	Frequency	
13	Period	
	Frequency	

Results and Conclusion:

Further Investigation Results:

Evaluation and Review Questions

1. Why is it important to check the dc voltage from a power supply *before* connecting it to a logic circuit?

2. Both analog and digital oscilloscopes have four major categories of controls. In your own words, explain the function of each section:
   a. vertical section

   b. trigger section

   c. horizontal section

   d. display section

3. Explain how to measure voltage across an ungrounded component with a two-channel oscilloscope.

4. In Step 11, you constructed a digital oscillator. Assume each of the following faults were in the circuit (each fault is independent of the others). Explain what symptom you would expect to observe with an oscilloscope.
   a. The LED is inserted in reverse.

   b. The value of C_1 is larger than it should be.

c. The power and ground connections on the power supply were accidentally reversed. (Don't test this one!)

d. R_1 is open.

5. Compare the advantage and disadvantage of making a dc voltage measurement with a DMM and a scope.

6. Explain how a logic pulser and a current tracer can be used to find a short between power and ground on a circuit board.

Experiment 2
Constructing a Logic Probe

Objectives

After completing this experiment, you will be able to

☐ Construct a simple logic probe using a 7404 inverter.

☐ Use this logic probe to test another circuit.

☐ Measure logic levels with the digital multimeter (DMM) and the oscilloscope, and compare them with valid input logic levels.

Materials Needed

7404 hex inverter
Two LEDs (light-emitting diodes)
Two signal diodes (1N914 or equivalent)
Resistors: three 330 Ω, one 2.0 kΩ
1 kΩ potentiometer

Summary of Theory

Digital circuits have two discrete voltage levels to represent the binary digits *(bits)* 1 and 0. All digital circuits are switching circuits that use high-speed transistors to represent either an ON condition or an OFF condition. Various types of logic, representing different technologies, are available to the logic designer. The choice of a particular family is determined by factors such as speed, cost, availability, noise immunity, and so forth. The key requirement within each family is compatibility; that is, there must be consistency within the logic levels and power supplies of various integrated circuits made by different manufacturers. The experiments in this lab book use primarily transistor-transistor logic, or TTL. The input logic levels for TTL are illustrated in Figure 2–1.

For any integrated circuit (IC) to function properly, power and ground must be connected. The connection diagram for the IC shows these connections, although in practice the power and ground connections are frequently omitted from diagrams of logic circuits. Figure 2–2 shows the connection diagram for a 7404 hex inverter, which will be used in this experiment.* Pins are numbered counter-clockwise from the top, starting with a notch or circle at the top or next to pin 1; see Figure 2–3.

The circuit in this experiment is a simple logic probe. Logic probes are useful for detecting the presence of a HIGH or a LOW logic level in a circuit. The logic probe in this experiment is designed only to illustrate the use of this tool and the wiring of integrated circuits. The probe shown in Figure 2–4 works as follows: If the probe is not connected, the top inverter is pulled HIGH (through the 2.0 kΩ resistor) and the bottom inverter is pulled LOW (through the 330 Ω resistor). As a result, both outputs are HIGH and neither LED is on. (A LOW is required to turn on either LED). If the probe input is connected to a voltage above approximately 2.0 V, the voltage at the input of the lower inverter is interpreted as a logic HIGH through diode D_2. As a

---

*Appendix A contains connection diagrams as found in manufacturers' logic data books and on their websites.

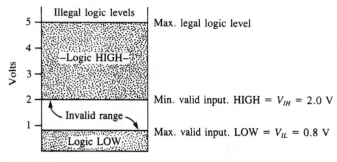

FIGURE 2–1
TTL logic levels.

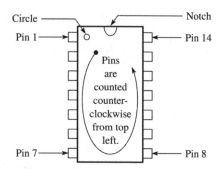

FIGURE 2–3
Numbering of pins.

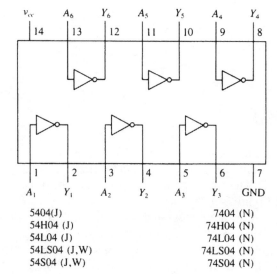

5404(J)	7404 (N)
54H04 (J)	74H04 (N)
54L04 (J)	74L04 (N)
54LS04 (J,W)	74LS04 (N)
54S04 (J,W)	74S04 (N)

FIGURE 2–2
Connection diagram.

result, the output of the lower inverter goes LOW, and the lower LED, representing a HIGH input, turns on. If the probe input is connected to a voltage below approximately 0.8 V, the upper input inverter is pulled below the logic threshold for a LOW, and the output inverter is LOW. Then the upper LED,

which represents a logic LOW input, turns on. A more sophisticated probe could detect pulses, have a much higher input impedance, and be useful for logic families other than TTL; however, this probe will allow you to troubleshoot basic gates.

Procedure

A Simple Logic Probe

1. Using the pin numbers shown, construct the simple logic probe circuit shown in Figure 2–4. Pin numbers are included on the drawing but frequently are omitted from logic drawings. Note that the LEDs and the signal diodes are polarized; that is, they must be connected in the correct direction. The arrow on electronic components always points in the direction of *conventional* current flow, defined as from plus to minus. Signal diodes are marked on the cathode side with a line. The LEDs generally have a flat spot on the cathode side or are the longer element inside the diode. As a guide, Figure 2–5(a) shows an example of the wiring of the logic probe.

2. Test your circuit by connecting the probe to +5.0 V and then to ground. One of the LEDs should light to indicate a HIGH, the other, a LOW. When

FIGURE 2–4
Simple logic probe.

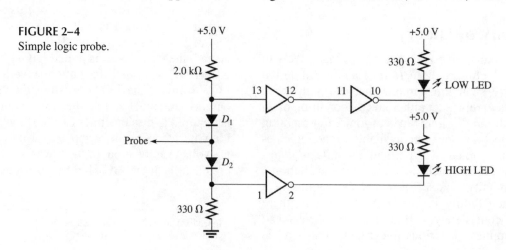

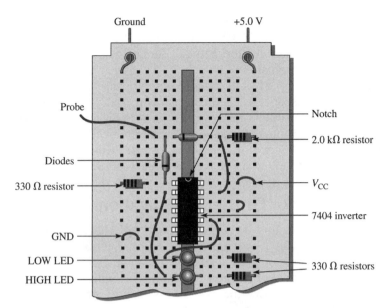

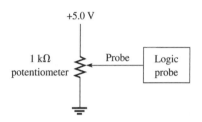

(a) Wiring of a logic probe

(b) Test circuit to determine logic thresholds

FIGURE 2–5

the probe is not connected, neither LED should be on. If the circuit does not work, double-check your wiring and the direction of the diodes.

3. You can test the HIGH and LOW threshold voltages of your logic probe with the circuit shown in Figure 2–5(b). Connect the logic probe to a 1 kΩ potentiometer, as shown. Vary the resistor and find the HIGH and LOW thresholds. Use a DMM to measure the threshold voltages. Record the thresholds in the report.

4. In the last step, you should have observed that the thresholds for the logic probe are very close to the TTL specifications given in Figure 2–1. If so, you can now use the probe to test the logic for logic gates including various inverter circuits. Remember that there are six independent inverters in a 7404 (but they share a common power supply). Begin by testing the inverter that is between pins 3 and 4 (this is inverter 2). Connect your logic probe to the output (pin 4) and observe the output logic when the input

is LOW (use ground), is OPEN, and is HIGH (use +5 V). Record your observations for these three cases in Table 2–1. An open input will have an invalid logic level; however, the output will be a *valid* logic level (an open input is not desirable because of potential noise problems).

5. Connect two inverters in series (cascade) as shown in Figure 2–6(a). Move the logic probe to the output of the second inverter (pin 6). Check the logic when the input is connected to a logic LOW, OPEN, and HIGH as before. Record your observations for these three cases in Table 2–1.

6. Connect the two inverters as cross-coupled inverters as shown in Figure 2–6(b). This is a basic latch circuit, the most basic form of memory. This arrangement is not the best way to implement a latch but serves to illustrate the concept (you will study latch circuits in more detail in Experiment 14). This latch works as follows: the input signal is first inverted by the top inverter. The original logic

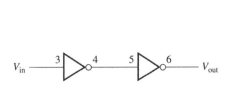

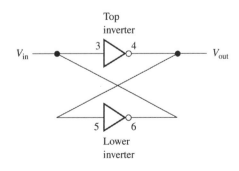

FIGURE 2–6　(a) Series inverters

(b) Cross-coupled inverters

is inverted a second time by the lower inverter which restores the logic back to the original input logic level (similar to your observation in step 5). This is a "feedback" signal which forms the latch. If the input is now removed, the feedback signal keeps the input from changing and the circuit remains stable. You will test this in the next step.

7. Connect the logic probe to pin 4 (the output) of the latch circuit. Then momentarily touch V_{in} (pin 3) to ground. Observe the output logic and record it (HIGH, LOW, or INVALID) in Table 2–2 of the report.

8. Touch the input to +5.0 V, test the output again, and record the logic in Table 2–2.

9. Place a fault in the circuit of Figure 2–6(b) by removing the wire that is connected to pin 5, the input of the lower inverter. Now momentarily touch the input, pin 3, to ground. Test the logic levels at each point in the circuit and record them in Table 2–2.

10. An open circuit on the input of TTL logic has an invalid logic level. Even though it is invalid, it acts as a logic HIGH at the input to the gate. (However, open circuits should never be used as a means of connecting an input to a constant HIGH.) Repeat Step 9 but use a DMM to measure the actual voltages at each pin. Record the data in Table 2–2.

11. In order to gain practice with the oscilloscope, repeat the measurements of Step 10 using the oscilloscope. You may want to review the procedure for making dc voltage measurements with the oscilloscope in the Oscilloscope Guide. Record the measured voltages in Table 2–2.

For Further Investigation

In this investigation, you will check the logic inverter with a pulse waveform. Set up the pulse generator with a 1 kHz TTL compatible pulse. Set up the series inverters shown in Figure 2–6(a). Then compare the waveforms on the input and on the output of the circuit. Sketch the waveforms in Plot 1 provided in the Report section; be sure and label the voltage and time. Are the waveforms identical? If not, why not? Explain your observations in the space provided in the Report section.

Report for Experiment 2

Name: _____ Date: _____ Class: _____

Objectives:

☐ Construct a simple logic probe using a 7404 inverter.
☐ Use this logic probe to test another circuit.
☐ Measure logic levels with the digital multimeter and the oscilloscope, and compare them with valid input logic levels.

Data and Observations:

Step 3: Logic thresholds: HIGH _____ V   LOW _____ V

TABLE 2–1

Step		Output Logic Level		
		Input is LOW	Input is OPEN	Input is HIGH
4	One inverter			
5	Two series inverters			

TABLE 2–2

Step		Input Logic Level (pin 3)	Output Logic Level (pin 4)	Logic Level (pin 5)	Logic Level (pin 6)
7	V_{in} momentarily on ground.				
8	V_{in} momentarily on +5.0 V.				
9	Fault condition: open at pin 5.				
10	Voltages with fault (DMM):	V	V	V	V
11	Voltages with fault (scope):	V	V	V	V

Results and Conclusion:

17

Further Investigation Results:

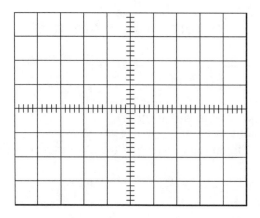

PLOT 1   Input and output waveforms
for two series inverters

Observations: _____

_____

Evaluation and Review Questions

1. In Step 3, you tested the threshold voltages of the logic probe. What simple change to the circuit of Figure 2–4 would you suggest if you wanted to raise these thresholds a small amount?

2. In Step 5, two inverters were connected in series; occasionally this configuration is used in logic circuits. What logical reason can you suggest that two inverters might be connected in series like this?

3. Consider the logic drawing in Figure 2–7.

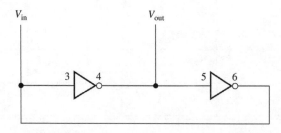

FIGURE 2–7

   a. Is this the same or different than the circuit in Figure 2–6(b)? _____

b. If the conductor leading to pin 3 is open, what voltage do you expect to see at pin 3? _____

c. If the conductor leading to pin 3 is open, what voltage do you expect to see at pin 4? _____

4. Discuss the advantage and disadvantage of using a logic probe and DMM for logic measurements.

5. Consider the circuit in Figure 2–8 with five inverters. Assume each inverter requires 10 ns for the input logic to affect the output logic (this is called *propagation delay*).

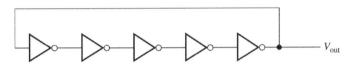

V_{out}

FIGURE 2–8

a. Describe V_{out}.

b. How long does it take for the input to the first inverter to affect the last inverter? _____

c. What is the frequency of the output? (Hint: The logic must change twice in one period.) _____

6. When troubleshooting a TTL logic circuit, what is the likely problem if a steady-state input voltage is invalid?

19

Experiment 3
Number Systems

Objectives

After completing this experiment, you will be able to
- Convert binary or binary coded decimal (BCD) numbers to decimal.
- Construct a portion of a digital system that decodes a BCD number and displays it on a seven-segment display.
- Troubleshoot the circuit for simulated faults.

Materials Needed

Four LEDs
7447A BCD/decimal decoder
MAN72 seven-segment display
Four-position DIP switch
Resistors: eleven 330 Ω, one 1.0 kΩ

For Further Investigation:
   Additional LED
   One 330 Ω resistor

Summary of Theory

The number of symbols in a number system is called the *base,* or *radix,* of that system. The decimal number system uses ten counting symbols, the digits 0 through 9, to represent quantities. Thus it is a base ten system. In this system, we represent quantities larger than 9 by using positional weighting of the digits. The position, or column, that a digit occupies indicates the weight of that digit in determining the value of the number. The base ten number system is a weighted system because each column has a value associated with it.

Digital systems use two states to represent quantities and thus are *binary* in nature. The binary counting system has a radix of two and uses only the digits 0 and 1. (These are often called *bits,* which is a contraction of BInary digiT). It too is a weighted counting system, with each column value worth twice the value of the column to the immediate right. Because binary numbers have only two digits, large numbers expressed in binary require a long string of 0s and 1s. Other systems, which are related to binary in a simple way, are often used to simplify these numbers. These systems include octal, hexadecimal, and BCD.

The octal number system is a weighted number system using the digits 0 through 7. The column values in octal are worth 8 times that of the column to the immediate right. You convert from binary to octal by arranging the binary number in groups of 3 bits, starting at the binary point, and writing the octal symbol for each binary group. You can reverse the procedure to convert from octal to binary. Simply write an equivalent 3-bit binary number for each octal character.

The hexadecimal system is a weighted number system using 16 characters. The column values in hexadecimal (or simply hex) are worth 16 times that of the column to the immediate right. The characters are the numbers 0 through 9 and the first six letters of the alphabet, A through F. Letters were chosen because of their sequence, but remember that they are used to indicate numbers, not letters. You convert binary numbers to hexadecimal numbers by arranging the binary number into 4-bit groups, starting at the binary point. Then write the next symbol for each group of 4 bits. You convert hex numbers to binary by reversing the procedure. That is, write an equivalent 4-bit binary number for each hexadecimal character.

The BCD system uses four binary bits to represent each decimal digit. It is a convenient code because it allows ready conversion from base ten to a code that a machine can understand; however, it is wasteful of bits. A 4-bit binary number could represent the numbers 0 to 15, but in BCD it represents only the quantities 0 through 9. The binary representations of the numbers 10 through 15 are not used in BCD and are invalid.

The conversion of BCD to a form that can be read by humans is a common problem in digital systems. A familiar display is called the seven-segment display, which is used in many digital applications such as clocks. A basic seven-segment display is described in this experiment with the instructions for wiring it. Later, in the text, the inner workings of the decoder and seven-segment display are explored. The block diagram for the tablet-bottling system that was introduced in Section 1–7 of the text is an example of a small system that used a BCD to seven-segment conversion circuit. It is used in both the on-site display and the remote display portions of that system. You will construct a simplified version of the display for this experiment.

Procedure

1. Take a moment to review "Circuit Wiring" in the Introduction before constructing the circuit in this experiment. The pin numbers for the integrated circuits (ICs) in this and succeeding experiments are not shown; pin numbers may be found on the data sheets in Appendix A or on the manufacturer's website. It is a good idea to write the pin numbers directly on the schematic before you begin wiring.

2. Begin by constructing the circuit shown in Figure 3–1, which will represent a BCD input. After wiring the circuit, connect power and test each switch to see that it lights an LED.

3. Remove power and add the circuit shown in Figure 3–2. An example of the wiring is shown in Figure 3–3. If you have not already done so, write the pin numbers on the schematic. The pin numbers for the MAN72 display are shown in Figure 3–4.*
Note that the 7447A has 16 pins, but the MAN72 has only 14 pins.

Before applying power, check that you have connected a 330 Ω current-limiting resistor between each output of the decoder and the input to the MAN72. Connect the Lamp test, BI/RBO, and RBI inputs through a 1.0 kΩ resistor to +5.0 V. This is a *pull-up resistor,* used to assure a solid logic HIGH is present at these inputs.

---

*Pin numbers for the 7447A decoder can be found in Appendix A.

FIGURE 3–1

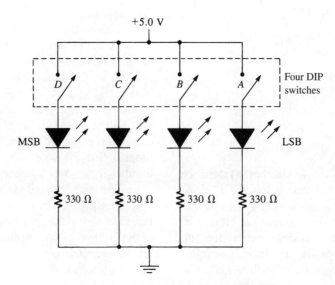

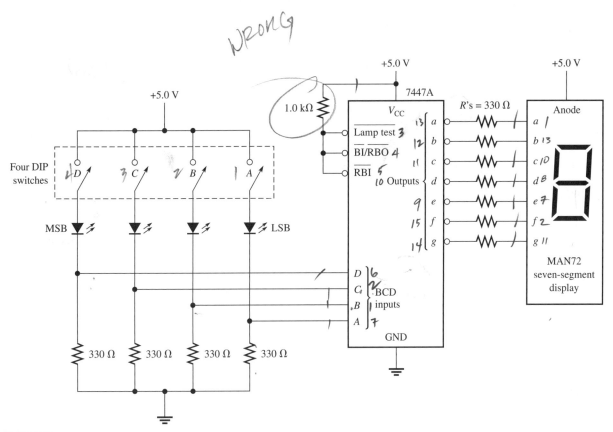

FIGURE 3–2

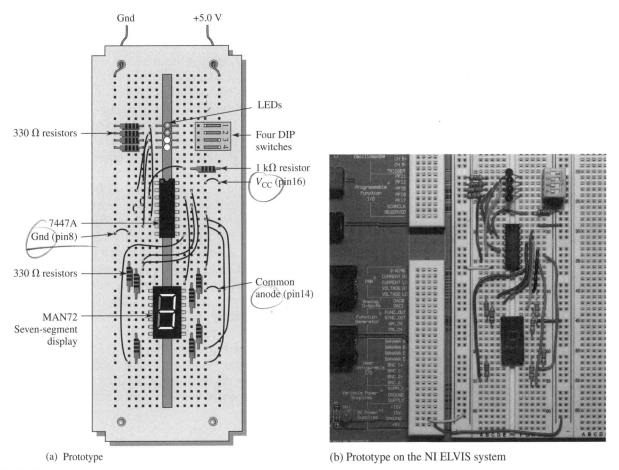

(a) Prototype

(b) Prototype on the NI ELVIS system

FIGURE 3–3

FIGURE 3–4
MAN72 seven-segment display.

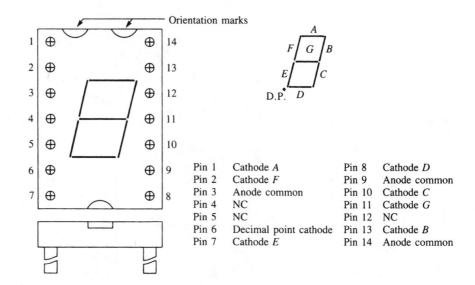

Orientation marks

Pin 1	Cathode *A*	Pin 8	Cathode *D*
Pin 2	Cathode *F*	Pin 9	Anode common
Pin 3	Anode common	Pin 10	Cathode *C*
Pin 4	NC	Pin 11	Cathode *G*
Pin 5	NC	Pin 12	NC
Pin 6	Decimal point cathode	Pin 13	Cathode *B*
Pin 7	Cathode *E*	Pin 14	Anode common

4. When you have completed the wiring, apply power, and test the circuit by setting each switch combination listed in Table 3–1 of the report. The last six codes are invalid BCD codes; however, you can set the switch combinations in binary and observe the display. It will show a unique display for each of the invalid codes. Complete the table by showing the appearance of the seven-segment display in the output column.

5. In this step, you will insert some simulated "troubles" in the circuit and observe the effect of these troubles on the output. The troubles are listed in Table 3–2 of the report. Insert the given trouble, and test its effect. Indicate what effect it has on the output. Assume each trouble is independent of others; that is, restore the circuit to its normal operating condition after each test.

6. The display you have built for this experiment is satisfactory only for showing a single decimal digit at a time. You could show more digits by simply replicating the circuit for as many digits as needed, although this isn't the most efficient way to make larger displays. With a larger number of digits, it is useful to blank (turn off) leading zeros in a number. Look at the function table in the manufacturer's specification sheet for the 7447A and decide what has to be applied to the $\overline{\text{Lamp test}}$, $\overline{\text{BI/RBO}}$, and $\overline{\text{RBI}}$ inputs in order to suppress leading zeros.* Summarize the method in the space provided in the report.

For Further Investigation

As you observed, the 7447A decoder used in this experiment is designed for BCD-to-decimal decoding; however, a slight modification of the circuit can be made to decode a binary number into octal. The largest number that we can show with a 4-bit binary input is octal 17, which requires two seven-segment displays to show both digits.

Recall that the conversion of a binary number to octal can be accomplished by grouping the binary number by threes, starting at the binary point. To display the octal numbers larger than binary 111 would normally require a second decoder and seven-segment display. For this problem, the most significant digit is either a zero or a one; therefore, we can dispense with the extra decoder and we could even use an ordinary LED to represent the most significant digit. The seven-segment display you have will still show the least significant digit. Modify the circuit in Figure 3–2 so that it correctly shows the octal numbers from 0 to 17. For example, if the switches are set to binary 1011, your circuit should light the LED representing the most significant digit and the seven-segment display should show a three.

A partial schematic is shown as Figure 3–5 in the report to help you get started. Complete the schematic, showing how to connect the circuit to show octal numbers. Construct the circuit, test it, and summarize how it works.

---

*A discussion of these inputs is given in the text in System Example 5–3 on page 250.

Report for Experiment 3

Name: _____ Date: _____ Class: _____

Objectives:

☐ Convert binary or BCD numbers to decimal.
☐ Construct a portion of a digital system that decodes a BCD number and displays it on a seven-segment display.
☐ Troubleshoot the circuit for simulated faults.

Data and Observations:

TABLE 3–1

Inputs		Output
Binary Number	BCD Number	Seven-Segment Display
0 0 0 0		
0 0 0 1		
0 0 1 0		
0 0 1 1		
0 1 0 0		
0 1 0 1		
0 1 1 0		
0 1 1 1		
1 0 0 0		
1 0 0 1		
1 0 1 0	INVALID	
1 0 1 1	INVALID	
1 1 0 0	INVALID	
1 1 0 1	INVALID	
1 1 1 0	INVALID	
1 1 1 1	INVALID	

TABLE 3–2

Trouble Number	Trouble	Observations
1	LED for the C input is open.	
2	A input to 7447A is open.	
3	$\overline{\text{LAMP TEST}}$ is shorted to ground.	
4	Resistor connected to pin 15 of the 7447A is open.	

Step 6. Method for causing leading zero suppression: _____

_____

Results and Conclusion:

Further Investigation Results:

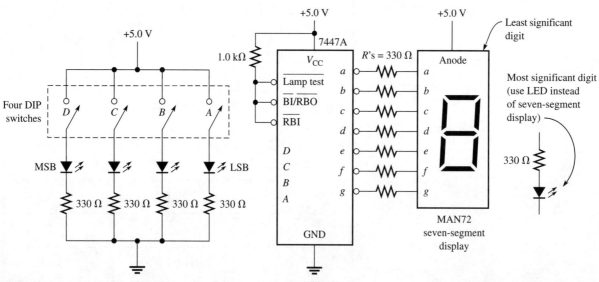

FIGURE 3–5

Evaluation and Review Questions

1. Assume the switches in Figure 3–2 are set for a binary 1000, but the display shows a zero. What are three possible causes for this error?

2. Looking at the possible causes for an error from Question 1, how would you go about troubleshooting the problem?

3. Suppose that the $\overline{\text{BI}}/\overline{\text{RBO}}$ input line was shorted to ground on the 7447A decoder and all other input lines were okay. Looking at the function table for the 7447A in Appendix A, determine the effect this would have on the display.

4. Explain the difference between binary and BCD.

5. Convert each number shown into the other bases:

Binary	Octal	Hexadecimal	Decimal	BCD
01001100				
	304			
		E6		
			57	
				0100 1001

6. a. The decimal number 85 is equal to 125 in a certain other number system. What is the base of the other system?

   b. The decimal number 341 is equal to 155 in a certain other number system. What is the base of the other system?

Experiment 4
Logic Gates

Objectives

After completing this experiment, you will be able to

☐ Determine experimentally the truth tables for the NAND, NOR, and inverter gates.

☐ Use NAND and NOR gates to formulate other basic logic gates.

☐ Use the ANSI/IEEE Std. 91–1984 logic symbols.

Materials Needed

7400 quad 2-input NAND gate
7402 quad 2-input NOR gate
Two 1.0 kΩ resistor

Summary of Theory

Logic deals with only two normal conditions: logic "1" or logic "0." These conditions are like the yes or no answers to a question. Either a switch is closed (1) or it isn't (0); either an event has occurred (1) or it hasn't (0); and so on. In Boolean logic, 1 and 0 represent conditions. In positive logic, 1 is represented by the term *HIGH* and 0 is represented by the term *LOW*. In positive logic, the more positive voltage is 1 and the less positive voltage is 0. Thus, for positive TTL logic, a voltage of $+2.4$ V $= 1$ and a voltage of $+0.4$ V $= 0$.

In some systems, this definition is reversed. With negative logic, the more positive voltage is 0 and the less positive voltage is 1. Thus, for negative TTL logic, a voltage of $+0.4$ V $= 1$ and a voltage of $+2.4$ V $= 0$.

Negative logic is sometimes used for emphasizing an active logic level. For all of the basic gates, there is a traditional symbol that is used for positive logic and an alternate symbol for negative logic. For example, an AND gate can be shown in negative logic with an OR symbol and "inverting bubbles" on the input and output, as illustrated with the three symbols in Figure 4–1(a). This logic can be read as "If A *or* B is LOW, the output is LOW." The exact same gate can be drawn as in Figure 4–1(b), where it is now shown as a traditional active-HIGH gate and read as "If both A *and* B are HIGH, the output is HIGH." The different drawings merely emphasize

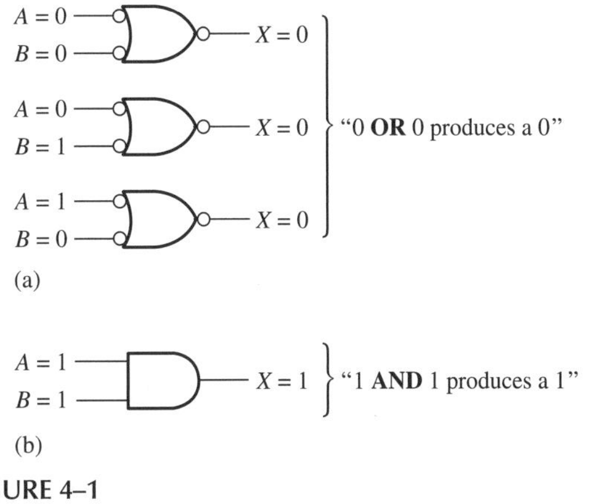

(a)

(b)

FIGURE 4–1

Two distinctive shape symbols for an AND gate. The two symbols represent the same gate.

one or the other of the following two rules for an AND gate:

> Rule 1: If A is LOW *or* B is LOW *or* both are LOW, then X is LOW.
>
> Rule 2: If A is HIGH *and* B is HIGH, then X is HIGH.

The operation can also be shown by the truth table. The AND truth table is

Inputs		Output	
A	*B*	*X*	
LOW	LOW	LOW	⎫
LOW	HIGH	LOW	⎬ Rule 1
HIGH	LOW	LOW	⎭
HIGH	HIGH	HIGH	← Rule 2

Notice that the first rule describes the first *three* lines of the truth table and the second rule describes the last line of the truth table. Although two rules are needed to specify completely the operation of the gate, each of the equivalent symbols best illustrates only one of the rules. If you are reading the symbol for a gate, read a bubble as a logic 0 and the absence of a bubble as a logic 1.

The first three lines of the truth table are illustrated with the negative-logic OR symbol (Figure 4–1(a)); the last line of the truth table is illustrated with the positive-logic AND symbol (Figure 4–1(b)). Similar rules and logic diagrams can be written for the other basic gates.

A useful method of dealing with negative logic is to label the signal function with a bar written over the label to indicate that the signal is LOW when the stated condition is true. Figure 4–2 shows some examples of this logic, called *assertion-level* logic. You should be aware that manufacturers are not always consistent in the way labels are applied to diagrams and function tables. Asser-

tion-level logic is frequently shown to indicate an action. As shown in Figure 4–2, the action to read (R) is asserted (1) when the input line is HIGH; the opposite action is to write ($\overline{W}$), which is asserted (0) when the line is LOW. Other examples are shown in the figure.

The symbols for the basic logic gates are shown in Figure 4–3. The newer ANSI/IEEE rectangular symbols are shown along with the older distinctive-shape symbols. The ANSI/IEEE symbols contain a qualifying symbol to indicate the type of logic operation performed. The distinctive-shape symbols for logic gates are still very popular because they enable you to visualize the standard Boolean operations of AND, OR, and INVERT immediately. The distinctive shapes also enable you to analyze logic networks because each gate can be represented with a positive logic symbol or an equivalent negative logic symbol. Both shapes are used in this experiment.

In addition to the AND, OR, and INVERT functions, two other basic gates are very important to logic designers. These are the NAND and NOR gates, in which the output of AND and OR, respectively, have been negated. These gates are important because of their "universal" property; they can be used to synthesize the other Boolean logic functions including AND, OR, and INVERT functions.

Two gates that are sometimes classified with the basic gates are the exclusive-OR (abbreviated XOR) and the exclusive-NOR (abbreviated XNOR) gates. These gates always have two inputs. The symbols are shown in Figure 4–3(f) and (g). The output of the XOR gate is HIGH when either A or B is HIGH, but not both (inputs "disagree"). The XNOR is just the opposite; the output is HIGH only when the inputs are the same (agree). For this reason, the XNOR gate is sometimes called a CO-INCIDENCE gate.

The logical operation of any gate can be summarized with a truth table, which shows all the possible inputs and outputs. The truth tables for INVERT, AND, OR, XOR, and XNOR are shown in

FIGURE 4–2
Examples of assertion-level logic.

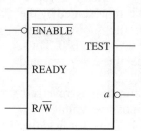

ENABLE is asserted (TRUE) when LOW.
READY is asserted (TRUE) when HIGH.
TEST is asserted (TRUE) when HIGH.
a is asserted (TRUE) when LOW.
R is asserted (TRUE) when HIGH.
$\overline{W}$ is asserted (TRUE) when LOW.

FIGURE 4–3
Basic logic gates.

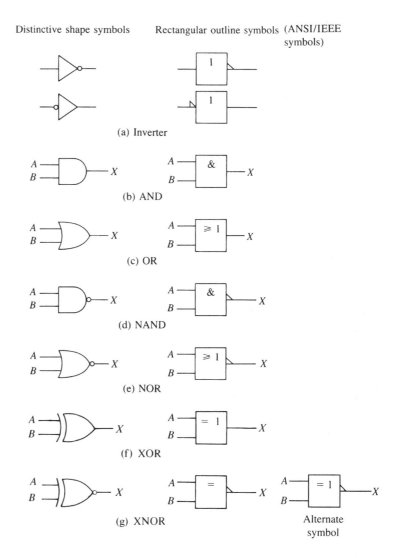

(a) Inverter

(b) AND

(c) OR

(d) NAND

(e) NOR

(f) XOR

(g) XNOR

Alternate symbol

Table 4–1(a) through (e). The tables are shown with 1 and 0 to represent positive logic HIGH and LOW, respectively. Except in Figure 4–1(a) (where negative logic is illustrated), only positive logic is used in this lab book and 1 and 0 mean HIGH and LOW, respectively.

In this experiment, you will test the truth tables for NAND and NOR gates as well as those for several combinations of these gates. Keep in mind that if any two truth tables are identical, then the logic circuits that they represent are equivalent. In the Further Investigation, look for this idea of equivalence between a 4-gate circuit and a simpler 1-gate equivalent.

Procedure

Logic Functions

1. Find the connection diagram for the 7400 quad 2-input NAND gate and the 7402 quad 2-input

NOR gate in the manufacturer's specification sheet.* Note that there are four gates on each of these ICs. Apply V_{CC} and ground to the appropriate pins. Then test one of the NAND gates by connecting all possible combinations of inputs, as listed in Table 4–2 of the report. Apply a logic 1 through a series 1.0 kΩ resistor and a logic 0 by connecting directly to ground. Show the logic output (1 or 0) as well as the measured output voltage in Table 4–2. Use the DMM to measure the output voltage.

2. Repeat Step 1 for one of the NOR gates; tabulate your results in Table 4–3 of the report.

3. Connect the circuits of Figures 4–4 and 4–5. Connect the input to a 0 and a 1, measure each output voltage, and complete truth Tables 4–4 and 4–5 for the circuits.

_____

*See Appendix A.

TABLE 4–1(a)
Truth table for inverter.

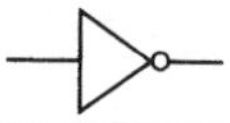

Input	Output
A	X
0	1
1	0

TABLE 4–1(b)
Truth table for 2-input AND gate.

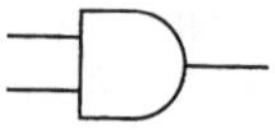

Inputs		Output
A	B	X
0	0	0
0	1	0
1	0	0
1	1	1

TABLE 4–1(c)
Truth table for 2-input OR gate.

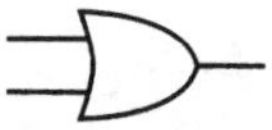

Inputs		Output
A	B	X
0	0	0
0	1	1
1	0	1
1	1	1

TABLE 4–1(d)
Truth table for XOR gate.

Inputs		Output
A	B	X
0	0	0
0	1	1
1	0	1
1	1	0

TABLE 4–1(e)
Truth table for XNOR gate.

Inputs		Output
A	B	X
0	0	1
0	1	0
1	0	0
1	1	1

4. Construct the circuit shown in Figure 4–6 and complete truth Table 4–6. This circuit may appear at first to have no application, but in fact can be used as a buffer. Because of amplification within the IC, a buffer provides more drive current.

5. Construct the circuit shown in Figure 4–7 and complete truth Table 4–7. Notice that the truth table for this circuit is the same as the truth table for one of the single gates. (What does this imply about the circuit?)

6. Repeat Step 5 for the circuits shown in Figures 4–8 and 4–9. Complete truth Tables 4–8 and 4–9.

For Further Investigation

The circuit shown in Figure 4–10 has the same truth table as one of the truth tables shown in Table 4–1(a) through (e). Test all input combinations and complete truth Table 4–10. What is the equivalent gate?

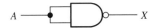

FIGURE 4–4

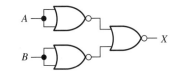

FIGURE 4–5

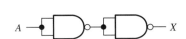

FIGURE 4–6

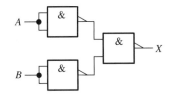

FIGURE 4–7

FIGURE 4–8

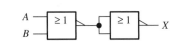

FIGURE 4–9

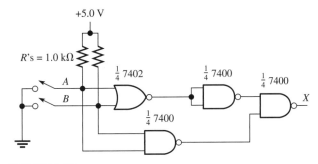

FIGURE 4–10

Report for Experiment 4

Name: _____ Date: _____ Class: _____

Objectives:

☐ Determine experimentally the truth tables for the NAND, NOR, and inverter gates.
☐ Use NAND and NOR gates to formulate other basic logic gates.
☐ Use the ANSI/IEEE Std. 91–1984 logic symbols.

Data and Observations:

TABLE 4–2
NAND gate.

Inputs		Output	Measured Output Voltage
A	B	X	
0	0		
0	1		
1	0		
1	1		

TABLE 4–3
NOR gate.

Inputs		Output	Measured Output Voltage
A	B	X	
0	0		
0	1		
1	0		
1	1		

TABLE 4–4
Truth table for Figure 4–4.

Input	Output	Measured Output Voltage
A	X	
0		
1		

TABLE 4–5
Truth table for Figure 4–5.

Input	Output	Measured Output Voltage
A	X	
0		
1		

TABLE 4–6
Truth table for Figure 4–6.

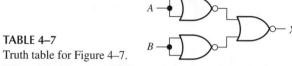

Input	Output	Measured Output Voltage
A	X	
0		
1		

TABLE 4–7
Truth table for Figure 4–7.

Inputs		Output	Measured Output Voltage
A	B	X	
0	0		
0	1		
1	0		
1	1		

TABLE 4–8
Truth table for Figure 4–8.

Inputs		Output	Measured Output Voltage
A	B	X	
0	0		
0	1		
1	0		
1	1		

TABLE 4–9
Truth table for Figure 4–9.

Inputs		Output	Measured Output Voltage
A	B	X	
0	0		
0	1		
1	0		
1	1		

Results and Conclusion:

Further Investigation Results:

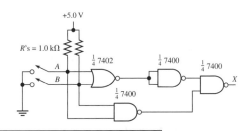

TABLE 4–10
Truth table for Figure 4–10.

Inputs		Output	Measured Output Voltage
A	B	X	
0	0		
0	1		
1	0		
1	1		

Evaluation and Review Questions

1. Look over the truth tables in your report.
   a. What circuits did you find that are equivalent to inverters?

   b. What circuit is equivalent to a 2-input AND gate?

   c. What circuit is equivalent to a 2-input OR gate?

2. An alarm circuit is needed to indicate that either the temperature *or* the pressure in a batch process is too high. If either of the conditions is true, a microswitch closes to ground, as shown in Figure 4–11. The required output for an LED is a LOW signal when the alarm condition is true. George thinks that an OR gate is needed, but Betty argues that an AND gate is needed. Who is right and why?

FIGURE 4–11

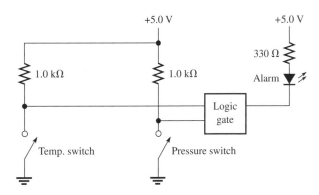

3. A car burglar alarm has a normally LOW switch on each of its four doors when they are closed. If any door is opened, the alarm is set off. The alarm requires an active-HIGH output. What type of basic gate is needed to provide this logic?

4. Suppose you needed a 2-input NOR gate for a circuit, but all you have available is a 7400 (quad 2-input NAND gate). Show how you could obtain the required NOR function using the NAND gate. (Remember that equivalent truth tables imply equivalent functions.)

5. A control signal that is used in a computer system is labeled $DT/\overline{R}$ for data transmit/receive. What action is implied when this signal is HIGH? LOW?

6. Assume you were troubleshooting a circuit containing a 4-input NAND gate and you discover that the output of the NAND gate is always HIGH. Is this an indication of a bad gate? Explain your answer.

Experiment 5
More Logic Gates

Objectives

After completing this experiment, you will be able to

☐ Determine experimentally the truth tables for OR and XOR.

☐ Test OR and XOR logic gates with pulse waveforms.

☐ Use OR and XOR gates to form a circuit that performs the 1's or 2's complement of a 4-bit binary number.

☐ Troubleshoot the complement circuit for simulated faults.

Materials Needed

ICs: one 7432 OR gate, one 7486 XOR gate
Eight LEDs
Resistors: nine 330 Ω, one 1.0 kΩ
One 4-position DIP switch
One SPST switch (wire may substitute)

For Further Investigation:
   Three additional 1.0 kΩ resistors

Summary of Theory

In this experiment, you will test the OR and XOR gates but go one step further and use these gates in an application.

The truth table for an OR gate is shown in Table 5–1(a) for a two-input OR gate. OR gates are available with more than two inputs. The operation of an *n*-input OR gate is summarized in the following rule:

The output is HIGH if any input is HIGH; otherwise it is LOW.

The XOR gate is a 2-input gate. Recall that the truth table is similar to the OR gate except for when both inputs are HIGH; in this case, the output is LOW. The truth table for a 2-input XOR gate can be summarized in the following statement:

The output is HIGH only if one input is HIGH; otherwise it is LOW.

The truth table for an XOR gate is shown in Table 5–1(b).

Procedure

Logic Functions for the OR and XOR Gates

1. Find the connection diagram for the 7432 quad 2-input OR gate and the 7486 quad 2-input XOR gate on the manufacturer's specification sheets.* Note that there are four gates on each of these ICs. Apply V_{CC} and ground to the appropriate pins. Then test one of the OR gates in the 7432 by

---

*See Appendix A.

TABLE 5–1(a)
Truth table for 2-input OR gate.

Inputs		Output
A	B	X
0	0	0
0	1	1
1	0	1
1	1	1

TABLE 5–1(b)
Truth table for XOR gate.

Inputs		Output
A	B	X
0	0	0
0	1	1
1	0	1
1	1	0

connecting all possible combinations of inputs, as listed in Table 5–2 of the report. Apply a logic 1 through a series 1.0 kΩ resistor and a logic 0 by connecting directly to ground. Show the logic output (1 or 0) as well as the measured output voltage in Table 5–2. Use the DMM to measure the output voltage.

2. Repeat Step 1 for one of the XOR gates in the 7486; tabulate your results in Table 5–3 of the report.

3. The XOR gate has a very useful feature enabling selective inversion of a waveform. Construct the circuit shown in Figure 5–1. The input on pin 2 is from your pulse generator, which should be set to a TTL compatible pulse. Set the frequency for 1 kHz and observe the input and the output simultaneously with S_1 open. Then close S_1 and observe the input and the output. Sketch the observed waveforms in Plot 1 of the report.

4. In this step, you will test a circuit that uses combinations of OR and XOR gates and complete the truth table for the circuit. The purpose of the circuit is to use the selective inversion property of the XOR gate to produce either the 1's complement or the 2's complement of a 4-bit number. Both the input and output numbers are read from the LEDs; the LEDs are on when the bit shown is LOW in keeping with TTL current specifications. Construct the circuit shown in Figure 5–2. You will need to assign pin numbers to the various pins.

5. Open the complement switch, and test the data switches. If the circuit is working properly, each output LED should be the exact opposite of the corresponding input LED. If this isn't what you observe, stop and troubleshoot your circuit.

6. Now test the circuit with the complement switch closed. Complete the truth table in the report (Table 5–4) for all possible inputs. Keep in mind that a 0 is indicated with an LED that is ON.

7. Table 5–5 (in the report) gives four possible problems that could occur in the complement circuit. For each problem given, list one or two likely causes that would produce the problem. As a check on your idea, you may want to test your idea on the circuit.

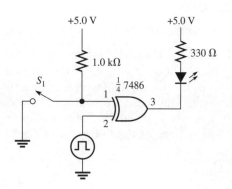

FIGURE 5–1

40

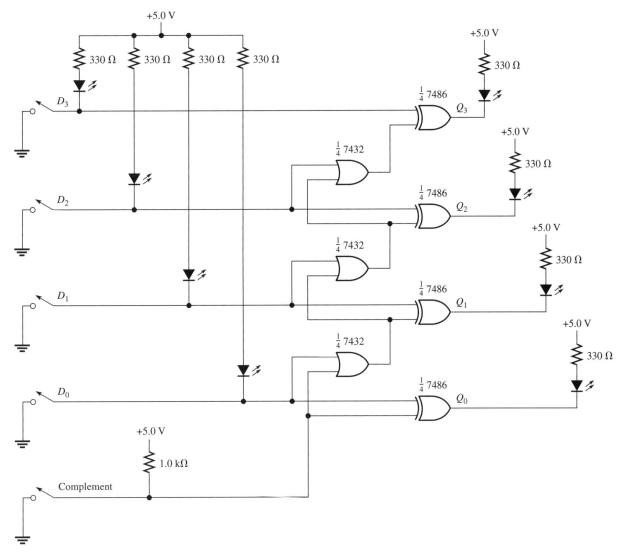

FIGURE 5–2

For Further Investigation

Another interesting circuit that can be constructed with XOR gates is the solution to the logic problem of controlling a light or other electrical device from several different locations. For two locations, the problem is simple and switches are made to do just that. The circuit shown in Figure 5–3 can control an LED from any of four locations. Construct and test the circuit; summarize your results in the report.

Multisim Troubleshooting (Optional)

The companion website* for this manual has Multisim 11 and 12 files for this experiment. Download the file named Exp-05nf (no-fault) and the worksheet Exp-05ws (worksheet). Open the file named Exp-05nf. This circuit is the same as in Figure 5–2, but pull-up resistors have been added to the data lines to allow the simulation to work correctly. Complete the worksheet and attach it to the report.

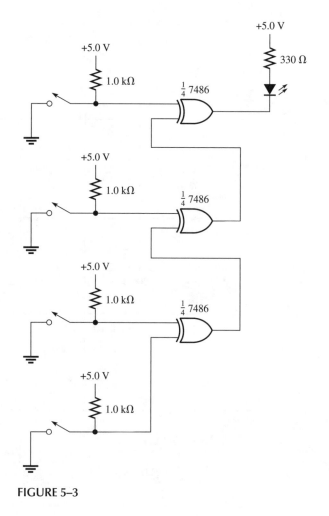

FIGURE 5–3

*www.pearsonhighered.com/floyd

Report for Experiment 5

Name: _____ Date: _____ Class: _____

Objectives:

☐ Determine experimentally the truth tables for OR and XOR.
☐ Test OR and XOR logic gates with pulse waveforms.
☐ Use OR and XOR gates to form a circuit that performs the 1's or 2's complement of a 4-bit binary number.
☐ Troubleshoot the complement circuit for simulated faults.

Data and Observations:

TABLE 5–2
OR gate.

Inputs		Output	Measured Output Voltage
A	B	X	
0	0		
0	1		
1	0		
1	1		

TABLE 5–3
XOR gate.

Inputs		Output	Measured Output Voltage
A	B	X	
0	0		
0	1		
1	0		
1	1		

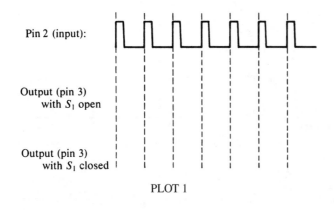

Pin 2 (input):

Output (pin 3)
with S_1 open

Output (pin 3)
with S_1 closed

PLOT 1

TABLE 5–4

Inputs	Outputs
$D_3\ D_2\ D_1\ D_0$	$Q_3\ Q_2\ Q_1\ Q_0$
0 0 0 0	
0 0 0 1	
0 0 1 0	
0 0 1 1	
0 1 0 0	
0 1 0 1	
0 1 1 0	
0 1 1 1	
1 0 0 0	
1 0 0 1	
1 0 1 0	
1 0 1 1	
1 1 0 0	
1 1 0 1	
1 1 1 0	
1 1 1 1	

TABLE 5–5

Symptom Number	Symptom	Possible Cause
1	None of the LEDs operate; the switches have no effect.	
2	LEDs on the output side do not work; those on the input side do work.	
3	The LED representing Q_3 is sometimes on when it should be off.	
4	The complement switch has no effect on the outputs.	

44

Results and Conclusion:

Further Investigation Results:

Evaluation and Review Questions

1. Step 3 mentions the selective inversion feature of an XOR gate. Explain how you can choose to invert or not invert a given signal.

2. The circuit in Figure 5–2 is limited to a 4-bit input. Show how you could expand the circuit to 8 bits by adding two more ICs.

3. The comparator in Figure 5–4 gives an output that depends on switches S_A, S_B, S_C, and S_D and inputs A, B, C, and D. Explain how the comparator works. (When is the output HIGH and when is it LOW?)

4. Redraw Figure 5–4 using the ANSI/IEEE Std. 91-1984 logic symbols.

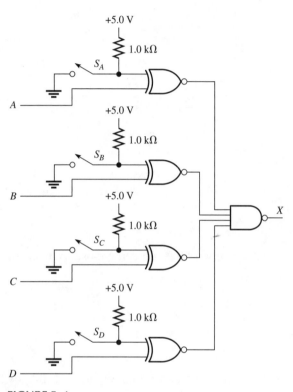

FIGURE 5–4

5. Assume you have two inputs, A and B, and their complements, $\overline{A}$ and $\overline{B}$, available. Show how you could use 2-input NAND gates to implement the XOR function.

6. Assume you have 2-input OR gates but needed to implement a 4-input OR function. Show how to connect the gates to implement the 4-input requirement.

Experiment 6
Interpreting Manufacturer's Data Sheets

Objectives

After completing this experiment, you will be able
to
□ Measure the static electrical specifications for
TTL and CMOS logic.
□ Interpret manufacturer's data sheets including
voltage and current requirements and limits.
□ Measure the transfer curve for a TTL inverter.

Materials Needed

7404 hex inverter
4081 quad AND gate
One 10 kΩ variable resistor
Resistors (one of each): 300 Ω, 1.0 kΩ, 15 kΩ,
1.0 MΩ, load resistor (to be calculated)

Summary of Theory

In this experiment, you will start by testing a TTL
(transistor-transistor logic) 7404 hex inverter, a
single package containing six inverters. As you
know, an inverter performs the NOT or complement
function. Ideally, the output is in either of two
defined states. As long as the manufacturer's specifications are not exceeded, these conditions will
hold and the logic will not be damaged.

TTL logic is designed to have conventional
current (plus to minus) flowing into the output terminal of the gate when the output is LOW. This
current, called *sink* current, is shown on a data sheet
as a *positive* current, indicating that it is flowing into

the gate. Conventional current leaves the gate when
the output is HIGH. This current is indicated on the
data sheet as a *negative* current and is said to be
source current. TTL logic can sink much larger
current than it can source.

In the last part of this experiment, you will test
CMOS (complementary metal-oxide semiconductor) logic. An important advantage to CMOS logic
is its low power consumption. When it is not switching from one logic level to another, the power dissipation in the IC approaches zero; however, at high
frequencies the power dissipation increases. Other
advantages include high fanout, high noise immunity, temperature stability, and ability to operate
from power supplies from 3 V to 15 V.

CMOS logic uses field-effect transistors,
whereas TTL uses bipolar transistors. This results in
significantly different characteristics. Consequently,
CMOS and TTL logic cannot be connected directly
to each other without due consideration of their
specifications, since voltage levels and current
sourcing and sinking capabilities differ. Interfacing
between various types of logic is determined by
these specifications. In addition, all MOS families
of logic are more sensitive to static electricity, and
special precautions to avoid static damage should be
observed when handling MOS devices. In addition
to static handling precautions, you should use the
following operating precautions:

1. Unused inputs must NOT be left open even on
   gates that are not being used. They should be tied
   to V_{CC}, ground, or an input signal.

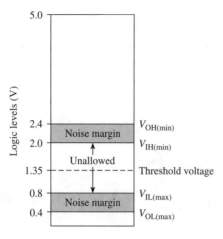

(a) TTL levels and noise margin

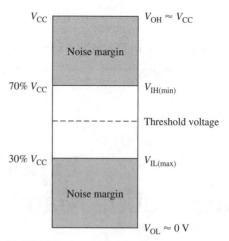

(b) CMOS levels and noise margin.

FIGURE 6-1

2. Power supply voltages must always be on when signal voltages are present at the inputs. Signal voltage must never exceed the power supply.
3. CMOS devices must never be inserted into or removed from circuits with the power on.

One important specification for any logic family is *noise margin*. Noise margin is the voltage difference that can exist between the output of one gate and the input of the next gate and still maintain guaranteed logic levels. For TTL, this difference is 0.4 V, as illustrated in Figure 6–1(a). For CMOS, the voltage thresholds are approximately 30% and 70% of V_{CC}. The noise margin is 30% of the supply voltage, as illustrated in Figure 6–1(b). A technique that is used to avoid noise problems in logic circuits is to place a small bypass capacitor (about 0.1 μF) between the supply and ground near every IC in a circuit.

A graphic tool to help visualize the characteristics of a circuit is called the *transfer curve*. The transfer curve is a graph of the input voltage plotted along the *x*-axis against the corresponding output voltage plotted along the *y*-axis. A linear circuit should have a straight-line transfer curve. On the other hand, a digital circuit has a transfer curve with a sharp transition between a HIGH and a LOW value. In the Further Investigation section, you will investigate the transfer curve for a 7404 inverter.

Procedure

1. The data table in the Report section is from the manufacturer's specified characteristics for the

7404 inverter. You will be measuring many of these characteristics in this experiment and entering your measured value next to the specified value. Begin by connecting the circuit shown in Figure 6–2(a). Inputs that are connected HIGH are normally connected through a resistor to protect the input from voltage surges and the possibility of shorting the power supply directly to ground in case the input is grounded.

2. Measure the voltage at the input of the inverter, with respect to ground, as shown in

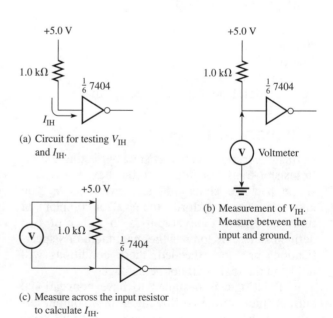

(a) Circuit for testing V_{IH} and I_{IH}.

(b) Measurement of V_{IH}. Measure between the input and ground.

(c) Measure across the input resistor to calculate I_{IH}.

FIGURE 6–2

Figure 6–2(b). Since the input is pulled HIGH, this voltage will be labeled V_{IH}. Enter your measured voltage in Table 6–1, line a. (Note that the specified V_{IH} is a *minimum* level; your measured value is probably much higher.)

3. Measure the voltage *across* the 1.0 kΩ resistor, as shown in Figure 6–2(c). TTL logic requires a very small input current to pull it HIGH. Using your measured voltage and Ohm's law, calculate the current in the resistor. This current is the input HIGH current, abbreviated I_{IH}. Enter your measured current in Table 6–1, line g. Compare your measured value with the specified maximum I_{IH}. Since conventional current is into the IC, the sign of this current is positive.

4. Measure the output voltage with respect to ground. Since the input is HIGH, the output is LOW. This voltage is called V_{OL}. Do not record this voltage; you will record V_{OL} in Step 5. Notice that without a load, the V_{OL} is much lower than the maximum specified level.

5. In order to determine the effect of a load on V_{OL}, look up the maximum LOW output current I_{OL} for the 7404. Then connect a resistor, R_{LOAD}, between the output and +5.0 V that allows $I_{OL(max)}$ to flow. Assume 4.8 V is dropped *across* the load resistor. Using Ohm's law, determine the appropriate load resistor, measure it, and place it in the circuit of Figure 6–3. Then measure V_{OL} and record it in Table 6–1, line f.

6. Measure the voltage drop across R_{LOAD}, and apply Ohm's law to determine the actual load current. Record the measured load current as I_{OL} in Table 6–1, line d.

7. Change the previous circuit to the one shown in Figure 6–4. Measure V_{IL} and V_{OH}, and record the measured voltages in Table 6–1, lines b and e.

8. Calculate I_{IL} by applying Ohm's law to the 300 Ω input resistor in Figure 6–4. (Note the sign.) Record the measured I_{IL} in Table 6–1, line h.

9. Measure the output load current by applying Ohm's law to the 15 kΩ load resistor. Record this current, I_{OH}, in Table 6–1, line c. Note the units and

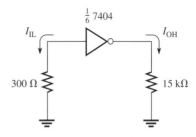

FIGURE 6–4

the sign. The minus sign indicates that the current is leaving the IC. This current, called *source* current, is significantly lower than the maximum LOW current, I_{OL}, which is positive current, or *sink* current.

10. In this and remaining steps, you will test a CMOS IC for several important characteristics. CMOS is static sensitive and it should be handled with special care. Avoid touching the pins. Always remove power before installing or removing a CMOS IC. Disconnect power from your protoboard and replace the 7404 with a 4081 quad AND gate. Although in practice you could test any CMOS gate, this gate will be used because it is needed in the next experiment and you have already tested an inverter. Check the manufacturer's specification sheet.* Enter the specified values for $V_{OL(max)}$, $V_{OH(min)}$, $V_{IL(max)}$, $V_{IH(min)}$, $I_{OL(min)}$, $I_{OH(min)}$ and $I_{IN(typ)}$ for a supply voltage of +5.0 V and a temperature of 25°C in Table 6–2. Notice on the pinout that the power is supplied to a pin labeled V_{DD} and ground is supplied to a pin labeled V_{SS}. Although this convention is commonly used, both pins are actually connected to transistor drains; V_{DD} is used to indicate the positive supply. (For +5.0 V supplies, the supply is often referred to as V_{CC}.)

11. Connect the circuit shown in Figure 6–5 and ground the inputs to all other gates except the two gates you are using. Reconnect the power supply and keep it at +5.0 V for these tests. Adjust the input voltage using the potentiometer to the manufacturer's

---

*See Appendix A.

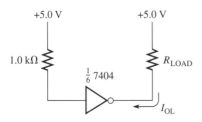

FIGURE 6–3

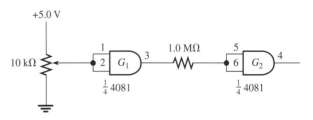

FIGURE 6-5

FIGURE 6–6

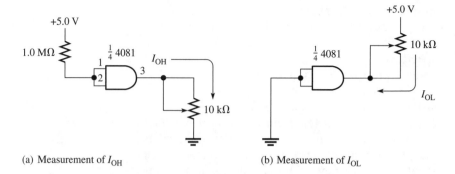

(a) Measurement of I_{OH} (b) Measurement of I_{OL}

specified value of $V_{IH(min)}$ for $V_{DD} = +5.0$ V. Record your measured value of the $V_{IH(min)}$ on line d of Table 6–2.

12. Read the output voltage of the first AND gate (G_1) at pin 3. Notice the difference between the CMOS gate and the TTL gate you tested earlier. Since the output is HIGH, record it as $V_{OH(min)}$ in Table 6–2, line b.

13. Measure the voltage *across* the 1.0 MΩ test resistor. A very high impedance meter is necessary to make an accurate measurement in this step. Determine the current flowing into the input of the second gate (G_2) by applying Ohm's law to the test resistor. Record this as the input current in Table 6–2, line g.

14. Adjust the potentiometer until the input voltage on G_1 is at the specified value of $V_{IL(max)}$. Record the measured input voltage in Table 6–2, line c. Measure the output voltage on pin 3 and record this as $V_{OL(max)}$ in Table 6–2, line a.

15. Turn off the power supply and change the circuit to that of Figure 6–6(a). The potentiometer is moved to the output and the 1.0 MΩ resistor is used as a pull-up resistor on the input. This circuit will be used to test the HIGH output current of the gate. After connecting the circuit, restore the power and adjust the potentiometer until the output voltage is 4.6 V (see manufacturer's stated conditions for specification of output current). Remove the power, measure the potentiometer's resistance, and apply Ohm's law to determine the output current I_{OH}. Record your measured current in Table 6–2, line f.

16. Change the circuit to that of Figure 6–6(b). Restore the power and adjust the potentiometer until the output voltage is 0.4 V. Remove

the power, measure the potentiometer's resistance, and apply Ohm's law to determine the output current I_{OL}. Remember, Ohm's law is applied by substituting the voltage measured *across* the resistor, not the output voltage. Record your measured current in Table 6–2, line e.

For Further Investigation

1. To investigate further the voltage characteristics of TTL, connect the circuit shown in Figure 6–7. The variable resistor is used to vary the input voltage.

2. Vary the input voltage through the range of values shown in Table 6–3. Set each input voltage; then measure the corresponding output voltage and record it in Table 6–3.

3. Plot the data from Table 6–3 onto Plot 1. Since V_{out} depends on V_{in}, plot V_{out} on the y-axis and V_{in} on the x-axis. This graph is called the *transfer curve* for the inverter.

4. Label the region on the transfer curve for V_{OH}, V_{OL}, and the threshold. The threshold is the region where the transition from LOW to HIGH takes place.

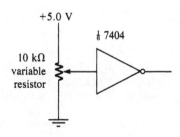

FIGURE 6–7

Report for Experiment 6

Name: _____ Date: _____ Class: _____

Objectives:

☐ Measure the static electrical specifications for TTL and CMOS logic.
☐ Interpret manufacturer's data sheets including voltage and current requirements and limits.
☐ Measure the transfer curve for a TTL inverter.

Data and Observations:

TABLE 6–1
TTL 7404.

Recommended Operating Conditions									
		DM5404			DM7404				
Symbol	Parameter	Min	Nom	Max	Min	Nom	Max	Units	Measured Value
V_{CC}	Supply voltage	4.5	5	5.5	4.75	5	5.25	V	
V_{IH}	High-level input voltage	2			2			V	a.
V_{IL}	Low-level input voltage			0.8			0.8	V	b.
I_{OH}	High-level output current			−0.4			−0.4	mA	c.
I_{OL}	Low-level output current			16			16	mA	d.
T_A	Free air operating temperature	−55		125	0		70	°C	

Electrical Characteristics Over Recommended Operating Free Air Temperature (unless otherwise noted)							
Symbol	Parameter	Conditions	Min	Typ	Max	Units	
V_I	Input clamp voltage	V_{CC} = Min, I_I = −12 mA			−1.5	V	
V_{OH}	High-level output voltage	V_{CC} = Min, I_{OH} = Max V_{IL} = Max	2.4	3.4		V	e.
V_{OL}	Low-level output voltage	V_{CC} = Min, I_{OL} = Max V_{IH} = Min		0.2	0.4	V	f.
I_I	Input current @ max input voltage	V_{CC} = Max, V_I = 5.5 V			1	mA	

Continued.

TABLE 6–1
Continued.

Electrical Characteristics Over Recommended Operating Free Air Temperature (unless otherwise noted)								
Symbol	Parameter	Conditions		Min	Typ	Max	Units	Measured Value
I_{IH}	High-level input current	V_{CC} = Max, V_I = 2.4 V				40	μA	g.
I_{IL}	Low-level input current	V_{CC} = Max, V_I = 0.4 V				−1.6	mA	h.
I_{OS}	Short circuit output current	V_{CC} = Max	DM54	−20		−55	mA	
			DM74	−18		−55		
I_{CCH}	Supply current with outputs HIGH	V_{CC} = Max			8	16	mA	
I_{CCL}	Supply current with outputs LOW	V_{CC} = Max			14	27	mA	

TABLE 6–2
CMOS CD4081.

	Quantity	Manufacturer's Specified Value	Measured Value
(a)	$V_{OL(max)}$, low-level output voltage		
(b)	$V_{OH(min)}$, high-level output voltage		
(c)	$V_{IL(max)}$, low-level input voltage		
(d)	$V_{IH(min)}$, high-level input voltage		
(e)	$I_{OL(min)}$, low-level output current		
(f)	$I_{OH(min)}$, high-level output current		
(g)	$I_{IN(typ)}$, input current		

Results and Conclusion:

Further Investigation Results:

TABLE 6–3

V_{in} (V)	V_{out} (V)
0.4	
0.8	
1.2	
1.3	
1.4	
1.5	
1.6	
2.0	
2.4	
2.8	
3.2	
3.6	
4.0	

PLOT 1

Evaluation and Review Questions

1. In Step 4, you observed V_{OL} with no load resistor. In Step 5, you measured V_{OL} with a load resistor. What is the effect of a load resistor on V_{OL}?

2. Assume a TTL logic gate has a logic HIGH output voltage of $+2.4$ V. Using the maximum specified I_{OH} from Table 6–1, determine the *smallest* output resistor that can be connected between the output and ground.

3. A hypothetical logic family has the following characteristics: $V_{IL} = +0.5$ V; $V_{IH} = +3.0$ V; $V_{OL} = +0.2$ V; $V_{OH} = +3.5$ V. Compute the LOW and HIGH noise margin for this family.

   V_{NL} (LOW) = _____; V_{NL} (HIGH) = _____

4. Assume that an LED requires 10 mA of current.
   a. Which of the two TTL circuits shown in Figure 6–8 is the better way to drive the LED? (Hint: look at I_{OL} and I_{OH}) _____. b. Why? _____

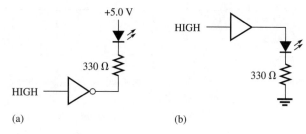

(a)   (b)

FIGURE 6–8

5. Explain why it is important for a troubleshooter to be aware of the ground level on an oscilloscope display.

6. Assume you connected the inputs of a TTL AND gate together and plotted the transfer curve. Sketch the shape of the curve and label the threshold.

Experiment 7
Boolean Laws and DeMorgan's Theorems

Objectives

After completing this experiment, you will be able to

- Experimentally verify several of the rules for Boolean algebra.
- Design circuits to prove Rules 10 and 11.
- Experimentally determine the truth tables for circuits with three input variables, and use DeMorgan's theorem to prove algebraically whether they are equivalent.

Materials Needed

4071 quad 2-input OR gate
4069 hex inverter
4081 quad 2-input AND gate
One LED
Four-position DIP switch
Four 1.0 kΩ resistors
Three 0.1 µF capacitors

Summary of Theory

Boolean algebra consists of a set of laws that govern logical relationships. Unlike ordinary algebra, where an unknown can take any value, the elements of Boolean algebra are binary variables and can have only one of two values: 1 or 0.

Symbols used in Boolean algebra include the overbar, which is the NOT or complement; the connective +, which implies logical addition and is read "OR"; and the connective ·, which implies logical multiplication and is read "AND." The dot is frequently eliminated when logical multiplication is shown. Thus $A \cdot B$ is written AB. The basic rules of Boolean algebra are listed in Table 7–1 for convenience.

The Boolean rules shown in Table 7–1 can be applied to actual circuits, as this experiment demonstrates. For example, Rule 1 states $A + 0 = A$ (remember to read + as "OR"). This rule can be demonstrated with an OR gate and a pulse generator, as shown in Figure 7–1. The signal from the pulse generator is labeled A and the ground represents the logic 0. The output, which is a replica of the pulse generator, represents the ORing of the two inputs—hence, the rule is proved. Figure 7–1 illustrates this rule.

In addition to the basic rules of Boolean algebra, there are two additional rules called DeMorgan's theorems that allow simplification of logic expressions that have an overbar covering more than one quantity. DeMorgan wrote two theorems for reducing these expressions. His first theorem is

The complement of two or more variables ANDed is equivalent to the OR of the complements of the individual variables.

Algebraically, this can be written as

$$\overline{X \cdot Y} = \overline{X} + \overline{Y}$$

His second theorem is

The complement of two or more variables ORed is equivalent to the AND of the complements of the individual variables.

Algebraically, this can be written as

$$\overline{X + Y} = \overline{X} \cdot \overline{Y}$$

As a memory aid for DeMorgan's theorems, some people like to use the rule "Break the bar and change the sign." The dot between ANDed quantities is implied if it is not shown, but it is given here to emphasize this idea.

The circuits constructed in this experiment use CMOS logic. You should use static protection to prevent damage to your ICs.

Procedure

1. Construct the circuit shown in Figure 7–1. Set the power supply to +5.0 V and use a 0.1 μF capacitor between V_{CC} and ground for each IC throughout this experiment.* If your pulse generator has a variable output, set it to a frequency of 10 kHz with a 0 to +4 V level on the output. Observe the signals from the pulse generator and the output at the same time on your oscilloscope. If you are using an analog scope, you need to be sure to trigger the scope on one channel only; otherwise a timing error can occur with some signals. The timing diagram and Boolean rule for this circuit has been completed in Table 7–2 in the report as an example.

2. Change the circuit from Step 1 to that of Figure 7–2. Complete the second line in Table 7–2.

3. Connect the circuit shown in Figure 7–3. Complete the third line in Table 7–2.

4. Change the circuit in Step 3 to that of Figure 7–4. Complete the last line in Table 7–2.

5. Design a circuit that will illustrate Rule 10. The pulse generator is used to represent the A input and a switch is used for the B input. Switch B is open for $B = 1$ and closed for $B = 0$. Complete the schematic, build your circuit, and draw two timing diagrams in the space provided in Table 7–3. The first timing diagram is for the condition $B = 0$ and the second is for the condition $B = 1$.

6. Design a circuit that illustrates Rule 11. Draw your schematic in the space provided in Table 7–4. Construct the circuit and draw two timing diagrams for the circuit in Table 7–4.

---

*In keeping with standard practice, capacitors are specified, particularly with CMOS devices, to return switching current "spikes" to the source through the shortest possible path.

56

TABLE 7–1
Basic rules of Boolean algebra.

1. $A + 0 = A$
2. $A + 1 = 1$
3. $A \cdot 0 = 0$
4. $A \cdot 1 = A$
5. $A + A = A$
6. $A + \overline{A} = 1$
7. $A \cdot A = A$
8. $A \cdot \overline{A} = 0$
9. $\overline{\overline{A}} = A$
10. $A + AB = A$
11. $A + \overline{A}B = A + B$
12. $(A + B)(A + C) = A + BC$

NOTE: A, B, or C can represent a single variable or a combination of variables.

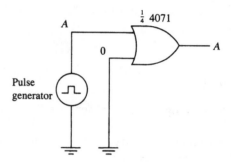

FIGURE 7–1

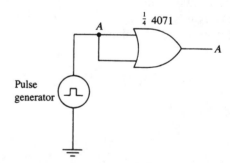

FIGURE 7–2

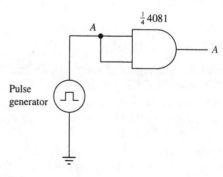

FIGURE 7–3

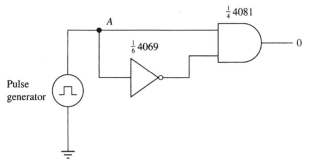

FIGURE 7–4

For Further Investigation

1. Build the circuit shown in Figure 7–5. Test each combination of input variables by closing the appropriate switches as listed in truth Table 7–5 in the report. Using the LED as a logic monitor, read the output logic, and complete Table 7–5.

2. Construct the circuit of Figure 7–6. Again, test each combination of inputs and complete truth Table 7–6 in the report. Compare the two truth tables. Can you prove (or disprove) that the circuits perform equivalent logic?

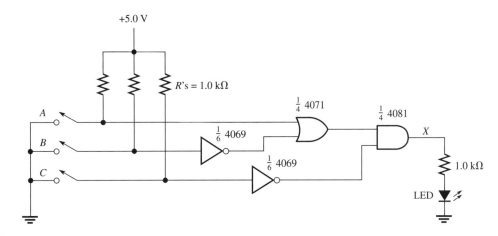

FIGURE 7–5

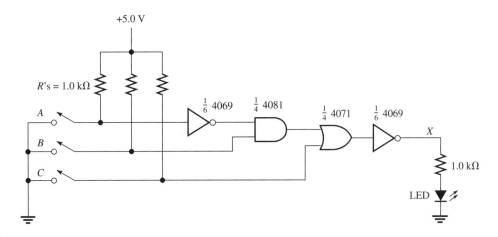

FIGURE 7–6

Report for Experiment 7

Name: _____ Date: _____ Class: _____

Objectives:

☐ Experimentally verify several of the rules for Boolean algebra.
☐ Design circuits to prove Rules 10 and 11.
☐ Experimentally determine the truth tables for circuits with three input variables, and use DeMorgan's theorem to prove algebraically whether they are equivalent.

Data and Observations:

TABLE 7–2

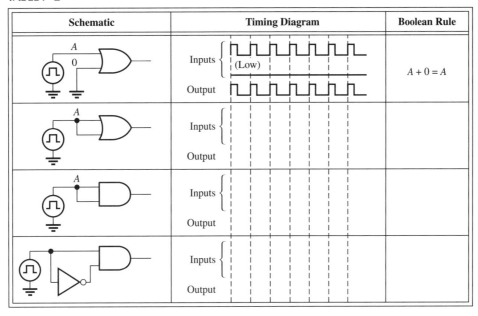

TABLE 7–3

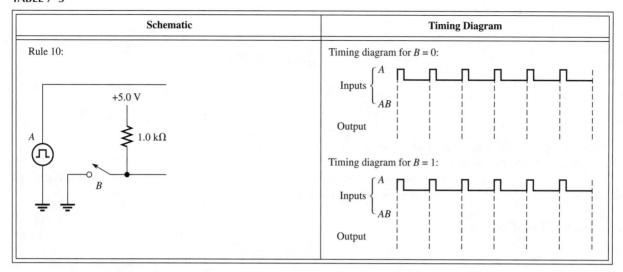

Schematic	Timing Diagram
Rule 10:	Timing diagram for $B = 0$:
	Timing diagram for $B = 1$:

TABLE 7–4

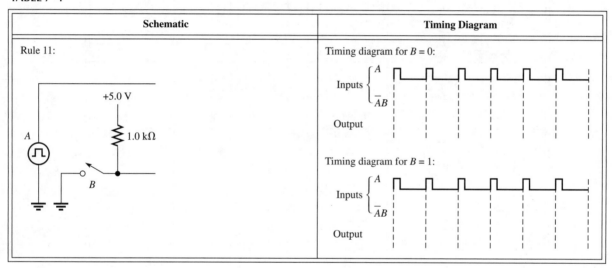

Schematic	Timing Diagram
Rule 11:	Timing diagram for $B = 0$:
	Timing diagram for $B = 1$:

Results and Conclusion:

Further Investigation Results:

TABLE 7–5
Truth table for Figure 7–5.

Inputs			Output
A	B	C	X
0	0	0	
0	0	1	
0	1	0	
0	1	1	
1	0	0	
1	0	1	
1	1	0	
1	1	1	

TABLE 7–6
Truth table for Figure 7–6.

Inputs			Output
A	B	C	X
0	0	0	
0	0	1	
0	1	0	
0	1	1	
1	0	0	
1	0	1	
1	1	0	
1	1	1	

Write the Boolean expression for each circuit.

Evaluation and Review Questions

1. The equation $X = A(A + B) + C$ is equivalent to $X = A + C$. Prove this with Boolean algebra.

2. Show how to implement the logic in Question 1 with NOR gates.

3. Draw two equivalent circuits that could prove Rule 12. Show the left side of the equation as one circuit and the right side as another circuit.

4. Determine whether the circuits in Figures 7–5 and 7–6 perform equivalent logic. Then, using DeMorgan's theorem, prove your answer.

5. Write the Boolean expression for the circuit shown in Figure 7–7. Then, using DeMorgan's theorem, prove that the circuit is equivalent to that shown in Figure 7–1.

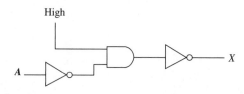

FIGURE 7–7

6. Assume the LED in Figure 7–5 is off no matter what the switch positions. List steps you would take to isolate the problem.

CHAPTER 4: COMBINATIONAL LOGIC

 # Experiment 8
Logic Circuit Simplification

Objectives

After completing this experiment, you will be able to

☐ Develop the truth table for a BCD invalid code detector.

☐ Use a Karnaugh map to simplify the expression.

☐ Build and test a circuit that implements the simplified expression.

☐ Predict the effect of "faults" in the circuit.

Materials Needed

7400 NAND gate
LED
Resistors: one 330 Ω, four 1.0 kΩ
One 4-position DIP switch

Summary of Theory

With combinational logic circuits, the outputs are determined solely by the inputs. For simple combinational circuits, truth tables are used to summarize all possible inputs and outputs; the truth table completely describes the desired operation of the circuit. The circuit may be realized by simplifying the expression for the output function as read from the truth table.

A powerful mapping technique for simplifying combinational logic circuits was developed by M. Karnaugh and was described in a paper he published in 1953. The method involved writing the truth table into a geometric map in which adjacent cells (squares) differ from each other in only one variable. (Adjacent cells share a common border horizontally or vertically.) When you are drawing a Karnaugh map, the variables are written in a Gray code sequence along the sides and tops of the map. Each cell on the map corresponds to one row of the truth table. The output variables appear as 0's and 1's on the map in positions corresponding to those given in the truth table.

As an example, consider the design of a 2-bit comparator. The inputs will be called A_2A_1 and B_2B_1. The desired output is HIGH if A_2A_1 is equal to or greater than B_2B_1. To begin, the desired output is written in the form of a truth table, as given in Table 8–1. All possible inputs are clearly identified by the truth table and the desired output for every possible input is given.

Next the Karnaugh map is drawn, as shown in Figure 8–1. In this example, the map is drawn using numbers to represent the inputs. The corresponding values for the output function are entered from the truth table. The map can be read in sum-of-products (SOP) form by grouping adjacent cells containing 1's on the map. The size of the groups must be an integer power of 2 (1, 2, 4, 8, etc.) and should contain only 1's. The largest possible group should be taken; all 1's must be in at least one group and may be taken in more than one group if helpful.

After grouping the 1's on the map, the output function can be determined. Each group is read as

TABLE 8–1
Truth table for comparator.

Inputs		Output
A_2 A_1	B_2 B_1	X
0 0	0 0	1
0 0	0 1	0
0 0	1 0	0
0 0	1 1	0
0 1	0 0	1
0 1	0 1	1
0 1	1 0	0
0 1	1 1	0
1 0	0 0	1
1 0	0 1	1
1 0	1 0	1
1 0	1 1	0
1 1	0 0	1
1 1	0 1	1
1 1	1 0	1
1 1	1 1	1

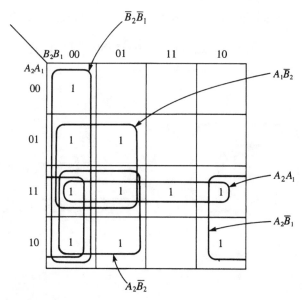

FIGURE 8–1
Karnaugh map for the truth table shown in Table 8–1.

one of the product terms in the reduced output function. Within each group larger than one, adjacent boundaries will be crossed, causing the variable that changes to be eliminated from the output expression. A group of two adjacent 1's will have a single adjacent boundary and will eliminate one variable. A group of four 1's will eliminate two variables and a group of eight 1's will eliminate three variables. Figure 8–1 shows the groupings for the 2-bit comparator. Since each group in this case is a group of four 1's, each product term contains two variables (two were eliminated from each term). The resulting expression is the sum of all of the product terms. The circuit can be drawn directly, as shown in Figure 8–2.

In this experiment, you will use the Karnaugh mapping method, similar to the one described previously, to design a BCD invalid code detector. As you know, BCD is a 4-bit binary code that represents the decimal numbers 0 through 9. The binary numbers 1010 through 1111 are invalid in BCD. You will design a circuit to assure that only valid BCD codes are present at the input and will signal a warning if an invalid BCD code is detected. Your circuit will be designed for 4 bits, but could easily be expanded to 8 bits.

Procedure

BCD Invalid Code Detector

1. Complete the truth table shown as Table 8–2 in the report. Assume the output for the ten valid BCD codes is a 0 and for the six invalid BCD codes is a 1. As usual for representing numbers, the most significant bit is represented by the letter D and the least significant bit by the letter A.

2. Complete the Karnaugh map shown as Figure 8–3 in the report. Group the 1's according to the rules given in the text and the Summary of Theory. Find the expression of the invalid codes by reading the minimum SOP from the map. Write the Boolean expression in the space provided in the report.

3. If you have correctly written the expression in Step 2, there are two product terms and you will see that the letter D appears in both terms. This expression could be implemented directly as a satisfactory logic circuit. By factoring D from each term, you will arrive at another expression for invalid codes. Write the new expression in the space provided in the report.

4. Recall that, for TTL logic, a LOW can light an LED without violating the I_{OL} (16 mA) specification but a HIGH causes the I_{OH} (400 µA) specification to be exceeded. To avoid this, the output is inverted and $\overline{X}$ is used to light the LED with a LOW logic level. The circuit shown in

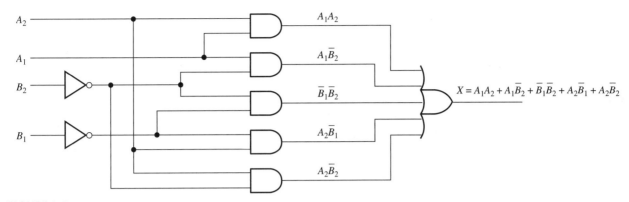

FIGURE 8–2
Circuit implementation of the comparator given by truth table in Table 8–1.

FIGURE 8–4

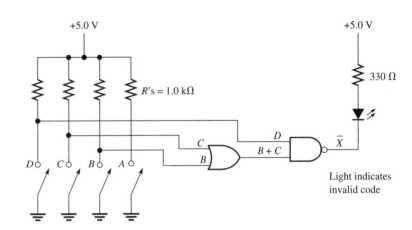

Figure 8–4 implements the expression from Step 3 but with the output inverted in order *sink* rather than *source* current.

5. Although the circuit shown in Figure 8–4 satisfies the design requirements with only two gates, it requires two *different* ICs. In some cases, this might be the best design. However, using the universal properties of the NAND gate, the OR gate could be replaced with three NAND gates. This change allows the circuit to be implemented with only *one* IC—a quad 7400. Change the circuit in Figure 8–4 by replacing the OR gate with three NAND gates. Draw the new circuit in the space provided in the report.

6. Construct the circuit you drew in Step 5. Test all combinations of the inputs and complete truth Table 8–3 in the report. If you have constructed and tested the circuit correctly, the truth table will be the same as Table 8–2.

7. For each problem listed in Table 8–4, state what effect it will have on your circuit. (Some "problems" may have no effect). If you aren't sure,

you may want to simulate the problem and test the result.

For Further Investigation

Design a circuit that will indicate if a BCD number is evenly divisible by three (3, 6, or 9). The input is a valid BCD number—assume that invalid numbers have already been tested and rejected by your earlier circuit! Since invalid numbers are now impossible, the Karnaugh map will contain "don't care" entries. An "X" on the map means that if the input is *not possible,* then you *don't care* about the outcome.

1. Complete the truth Table 8–5 in the report for the problem stated above. Enter 0's for BCD numbers that are not divisible by three and 1's for BCD numbers that are divisible by three. Enter an "X" to indicate an invalid BCD code.

2. Complete the Karnaugh map shown as Figure 8–5 in the report. Group the 1's on the map in groups of 2, 4, 8, etc. Do not take any 0's but do take

X's if you can obtain a larger group. Read the minimum SOP from the map and show your expression in the space provided in the report.

    3. Draw a circuit using only NAND gates that will implement the expression. The LED should be turned ON with a LOW output. Build the circuit and test each of the possible inputs to see that it performs as expected.

Multisim Troubleshooting (Optional)

The companion website* for this manual has Multisim 11 and 12 files. Download the file named Exp-08nf and the worksheet Exp-08ws. Open the file named Exp-08nf. It represents the circuit in Figure 8–4 and has no fault. It can be used as a reference. Complete the worksheet and attach it to the report.

---

*www.pearsonhighered.com/floyd

Report for Experiment 8

Name: _____ Date: _____ Class: _____

Objectives:

☐ Develop the truth table for a BCD invalid code detector.
☐ Use a Karnaugh map to simplify the expression.
☐ Build and test a circuit that implements the simplified expression.
☐ Predict the effect of "faults" in the circuit.

Data and Observations:

TABLE 8–2
Truth table for BCD invalid code detector.

Inputs				Output
D	C	B	A	X
0	0	0	0	
0	0	0	1	
0	0	1	0	
0	0	1	1	
0	1	0	0	
0	1	0	1	
0	1	1	0	
0	1	1	1	
1	0	0	0	
1	0	0	1	
1	0	1	0	
1	0	1	1	
1	1	0	0	
1	1	0	1	
1	1	1	0	
1	1	1	1	

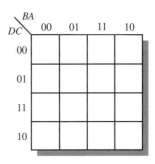

FIGURE 8–3
Karnaugh map of truth table for BCD invalid code detector.

Minimum sum-of-products read from map:

$X =$ _____

Factoring D from both product terms gives:

$X =$ _____

Step 5. Circuit for BCD invalid code detector:

TABLE 8–3

Truth table for BCD invalid code detector constructed in step 6.

Inputs				Output
D	C	B	A	X
0	0	0	0	
0	0	0	1	
0	0	1	0	
0	0	1	1	
0	1	0	0	
0	1	0	1	
0	1	1	0	
0	1	1	1	
1	0	0	0	
1	0	0	1	
1	0	1	0	
1	0	1	1	
1	1	0	0	
1	1	0	1	
1	1	1	0	
1	1	1	1	

TABLE 8–4

Problem Number	Problem	Effect
1	The pull-up resistor for the D switch is open.	
2	The ground to the NAND gate in Figure 8–4 is open.	
3	A 3.3 kΩ resistor was accidently used in place of the 330 Ω resistor.	
4	The LED was inserted backward.	
5	Switch A is shorted to ground.	

Results and Conclusion:

Further Investigation Results:

TABLE 8–5
Truth table for BCD numbers divisible by three.

Inputs				Output
D	C	B	A	X
0	0	0	0	
0	0	0	1	
0	0	1	0	
0	0	1	1	
0	1	0	0	
0	1	0	1	
0	1	1	0	
0	1	1	1	
1	0	0	0	
1	0	0	1	
1	0	1	0	
1	0	1	1	
1	1	0	0	
1	1	0	1	
1	1	1	0	
1	1	1	1	

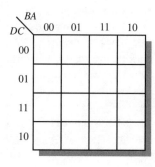

FIGURE 8–5
Karnaugh map of truth table for BCD numbers divisible by three.

Minimum sum-of-products read from map:

$X =$ _____

Circuit:

70

Evaluation and Review Questions

1. Assume that the circuit in Figure 8–4 was constructed but doesn't work correctly. The output is correct for all inputs except $DCBA = 1000$ and 1001. Suggest at least two possible problems that could account for this and explain how you would isolate the exact problem.

2. Draw the equivalent circuit in Figure 8–4 using only NOR gates.

3. The A input was used in the truth table for the BCD invalid code detector (Table 8–2) but was not connected in the circuit in Figure 8–4. Explain why not.

4. The circuit shown in Figure 8–6 has an output labeled $\overline{X}$. Write the expression for $\overline{X}$; then, using DeMorgan's theorem, find the expression for X.

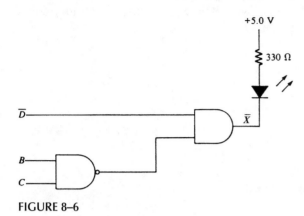

FIGURE 8–6

5. Convert the SOP form of the expression for the invalid code detector (Step 2) into POS form.

6. Draw a circuit, using NAND gates, that implements the invalid code detector from the expression you found in Step 2.

Experiment 9
The Perfect Pencil Machine

Objectives

After completing this experiment, you will be able to

☐ Design and build the combinational portion of a logic circuit that delivers a "pencil" and "change" based on the value of "coins" that are entered.

☐ Write a formal laboratory report describing your circuit and results.

Materials Needed

Four LEDs
One 4-position DIP switch
Resistors: four 1.0 kΩ, four 330 Ω
Other materials as determined by student

Summary of Theory

Most digital logic circuits require more than simple combinational logic. They require the ability to respond to a series of events. For example, a coin-operated machine must "remember" the amount of money that has been inserted previously and compare it with the price of the product. When the amount of money is equal or greater than the price of the product, the machine delivers the product and any necessary change. This is an example of a sequential machine; the key difference between it and a combinational machine is memory. Any circuit that contains memory is considered to be a sequential circuit.

For analysis purposes, sequential circuits can be broken into memory units and combinational units. A coin-operated machine actually contains both sequential (memory) and combinational portions. For simplification, we will focus only on the combinational logic in this experiment. You will complete the design of the combinational logic portion of a coin-operated machine that has multiple outputs. The outputs consist of a product and various coins for change.

To design the combinational logic portion of the machine, you will need a separate Karnaugh map for each output. Keep in mind that logic needed for one of the outputs sometimes appears in the equation for the other output, and this logic may simplify the total circuit.

As an example, the Model-1 perfect pencil machine design is shown (see Figure 9–1). This machine (which was built shortly after the invention of the pencil) uses combinational logic to determine if the mechanical hand holding the pencil should open. Back then, pencils sold for ten cents and the original pencil machine could accept either two nickels or one dime but nothing else. If someone first dropped in a nickel and then a dime, the machine was smart enough to give a nickel change.

The Model-1 perfect pencil machine was designed by first filling out a truth table showing all possibilities for the input coins. There were two switches for nickels, labeled N_1 and N_2, and one switch for a dime, labeled D. The truth table is shown as Table 9–1. The truth table lists two outputs; a pencil labeled P, and a nickel change labeled *NC*. A 1 on the truth table indicates that the coin has been placed in the machine or a product (pencil or change) is to be delivered; a 0 indicates the absence of a coin or that no product is to be delivered.

In the Model-1 machine, the two nickel switches are stacked on top of each other (see Figure 9–1), forming a mechanical version of sequential logic. It is *not possible* for the second nickel to be placed in the machine until the first nickel has been placed in the machine. The coins must be placed in the machine sequentially; as soon as ten cents is added, the pencil is delivered. It is not possible for three coins to be placed in the machine as any combination of two coins will deliver a pencil. For these reasons, several lines on the truth table are shown as "don't care" (X). "Don't cares" appear on the table because if the input is *not possible,* then we *don't care* what the output does.

The information from the truth table was entered onto two Karnaugh maps—one for each output (see Figure 9–2). The maps were read by taking "don't cares" wherever it would help. This, of course, greatly simplifies the resulting logic. The Boolean expressions that resulted surprised no one—not even the company president! From the Boolean sum-of-products (SOP), it was a simple

TABLE 9–1
Truth table for the Model-1 perfect pencil machine.

Inputs			Outputs	
N_1	N_2	D	P	NC
0	0	0	0	0
0	0	1	1	0
0	1	0	X	X
0	1	1	X	X
1	0	0	0	0
1	0	1	1	1
1	1	0	1	0
1	1	1	X	X

matter to implement the Model-1 perfect pencil machine, as shown in Figure 9–3.

Procedure

The logic for the new Model-2 perfect pencil machine must be designed and ready for testing by next week. Your job is to design the combinational logic portion of the new machine and test it (using LEDs for the outputs) to assure that it works. You

FIGURE 9–1
Model-1 perfect pencil machine.

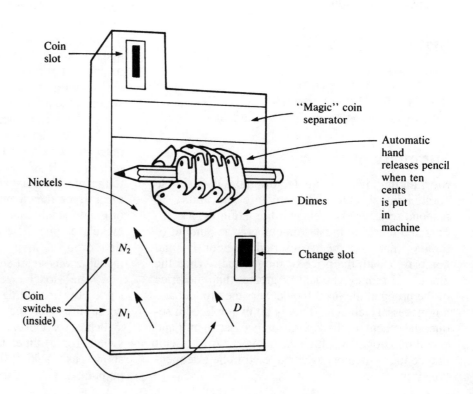

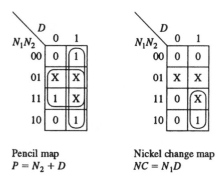

Pencil map
$P = N_2 + D$

Nickel change map
$NC = N_1 D$

FIGURE 9–2
Karnaugh maps for Model-1 perfect pencil machine.

must then write a report for the director of engineering summarizing your design and test results. The problem statement is given as follows.

Problem Statement: The new Model-2 perfect pencil machine sells pencils for fifteen cents (due to inflation). The machine will be designed to accept one nickel, two dimes, a quarter, or certain combinations of these. (The president of the company was heard to say that anyone who put three nickels in the machine deserved to lose them!). The coins for the combinational logic are represented by four input switches—a nickel (N), two dimes (D_1 and D_2), and a quarter (Q). The first dime in the machine will always turn on the D_1 switch and the

second dime in the machine will always turn on the D_2 switch. As in the Model-1 design, there are certain combinations that are impossible. This is due to the fact that the automatic hand (holding the pencil) begins to open as soon as 15 cents is in the machine. This implies that no more than 2 coins can be put in the machine. Also, it is not possible to activate the second dime switch before the first dime switch. To clarify the combinations, the director of engineering has started filling out the truth table (see Table 9–2) but was called to a meeting.

There are four outputs from the Model-2 perfect pencil machine, as listed on the truth table. They are pencil (P), nickel change (NC), the first dime change (DC_1), and the second dime change (DC_2). You will need to determine what these outputs should be to deliver the correct product and change.

Oh!—and the director of engineering says she would like to have you use no more than two ICs! Good luck!

Report

Write a technical report summarizing your circuit and your results of testing it. Your instructor may have a required format, or you may use the one given in the Introduction to the Student. Include the completed truth table for the model-2 Perfect Pencil machine and the Karnaugh maps showing the simplification you used.

For Further Investigation

After the Model-2 perfect pencil machine was designed, the purchasing department found out that 2-input NAND gates were on sale and they purchased several million of them. The problem is that you will need to modify your circuit to use the 2-input NAND gates throughout. Your boss muttered, "It may mean more ICs, but we've got to use up those NAND gates!" Show in the report how you will help the company in your design.

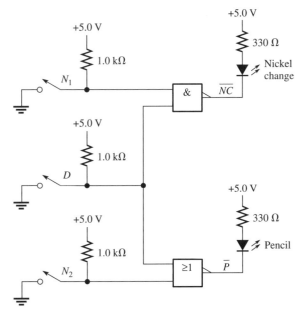

FIGURE 9–3
Model-1 perfect pencil machine.

TABLE 9–2
Truth table for Model-2 perfect pencil machine.

Inputs				Outputs			
N_1	D_1	D_2	Q	P	NC	DC_1	DC_2
0	0	0	0	0	0	0	0
0	0	0	1	1	0	1	0
0	0	1	0	X	X	X	X
0	0	1	1	X	X	X	X
0	1	0	0				
0	1	0	1				
0	1	1	0				
0	1	1	1				
1	0	0	0				
1	0	0	1				
1	0	1	0				
1	0	1	1				
1	1	0	0				
1	1	0	1				
1	1	1	0				
1	1	1	1				

Experiment 10
The Molasses Tank

Objectives

After completing this experiment, you will be able to

☐ Design and implement the logic for a control process that involves feedback.

☐ Write a formal laboratory report describing your circuit and results.

Materials Needed

Four LEDs
One four-position DIP switch
Resistors: four 1.0 kΩ, four 330 Ω
Other materials as determined by the student

Summary of Theory

Control systems typically use sophisticated programmable controllers to provide flexibility in implementing logic and allowing for changes in the design as requirements change. For simple systems, the control logic can be designed from fixed function logic, as will be the case in this experiment.

The problem posed in this experiment is to design the control logic for two of the four outputs in a tank control system. (The For Further Investigation section adds a third output, which is the inlet valve, V_{IN}). The specific requirements are described in the Problem Statement. In this experiment there is feedback present. Strictly speaking this means it

is sequential logic, but the methods employed in the design use combinational logic design methods (Karnaugh maps) like those discussed in Chapter 5. The feedback prevents the tank from refilling until it has emptied below a certain level as you will see. A flow sensor is not needed in this design.

The Crumbly Cookie Company has a problem with the storage tank for its new line of molasses cookies. The problem is that the molasses in the winter months runs too slow for the batch process. As a new employee, you have been assigned to design logic for the model-2 tank controller that will assure that the molasses is warm enough before the outlet valve, V_{OUT}, is opened. After it is opened, it must remain open until the lower sensor is uncovered, indicating the tank is empty.

The best way to understand the problem is to review the model-1 tank controller design. In the model-1 design, the molasses tank had two level sensors, one high, L_H, and one low, L_L. The tank was emptied (outlet valve opened) only when molasses reached the upper sensor. After opening, the valve was closed only when the lower sensor was uncovered.

As mentioned, the model-1 system opens the outlet valve *only* when both sensors are covered but, once opened, it remains open until both sensors are uncovered. This concept requires knowing the current state of the output valve; hence in the design, it is considered as both an *output* and an *input* to the logic. This idea is summarized with the truth table

TABLE 10–1

Truth table for the model-1 outlet valve. Note that V_{OUT} is both an input and an output becasue of the feedback.

Inputs			Output	
L_H	L_L	V_{OUT}	V_{OUT}	Action
0	0	0	1	Close valve.
0	0	1	1	Leave valve closed.
0	1	0	0	Valve is open; leave open.
0	1	1	1	Valve is closed; leave closed.
1	0	0	0	Sensor error; open valve.
1	0	1	0	Sensor error; open valve.
1	1	0	0	Sensors covered; leave valve open.
1	1	1	0	Sensors covered; open valve.

shown as Table 10–1. Because the system is designed for TTL logic, the outlet valve is *opened* with a LOW signal.

A Karnaugh map that represents the truth table is shown in Figure 10–1. From the map, the minimum logic is determined. The circuit for the outlet valve of the model-1 tank controller is shown in Figure 10–2. Notice that V_{OUT} is returned to one of the inputs, which is the feedback referred to previously.

After constructing the circuit for the outlet valve of the model-1 tank controller, it was tested using switches for the inputs and an LED to represent the outlet valve. The test begins with both switches closed, meaning the level inputs are both LOW. As the tank fills, L_L is covered, so it will open and changes to a HIGH level. The LED, representing the valve, remains OFF. Later, L_H is covered, so it is opened and changes to a HIGH level. This causes the LED to turn ON. At this point, closing the HIGH level switch will have no effect on the LED as it remains ON until L_L is again covered, represented by a LOW.

Procedure

Design the logic for the outlet valve, V_{OUT}, and the alarm, A, for the model-2 molasses tank controller. Test the design and submit a report showing the schematic and test results for this portion of the controller. The design requirements are given with the problem statement.

Problem Statement: The model-2 molasses tank controller has three inputs and four outputs, as shown in Figure 10–3. The inputs are two level sensors, L_L and L_H, and a temperature sensor, T_C. If the temperature is too cold for the molasses to flow properly, the "cold" temperature sensor will indicate this with a logic 1. The outputs are two valves, V_{IN}, and V_{OUT},

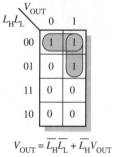

$$V_{OUT} = \overline{L_H}\,\overline{L_L} + \overline{L_H}V_{OUT}$$

FIGURE 10–1

Karnaugh map simplification for the outlet valve of the model-1 tank controller.

FIGURE 10–2

Schematic for the outlet valve logic of the model-1 tank controller.

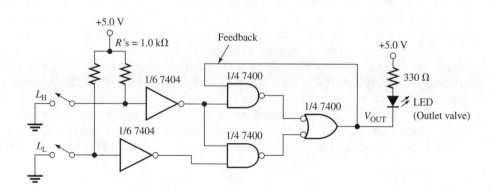

78

FIGURE 10–3
Model-2 tank controller.

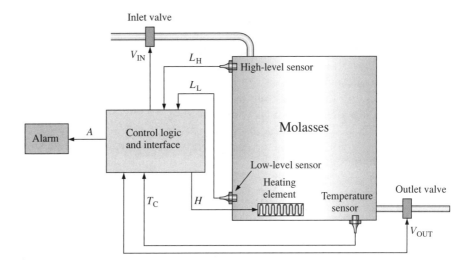

an alarm, A, and a heater, H, to warm the molasses sufficiently for it to flow properly. You need to design only the logic for the outlet valve, V_{OUT}, and the alarm, A. As an example, the truth table for V_{OUT} has been prepared (with comments) as Table 10–2.

As in the model-1 controller, when the upper level sensor indicates it is covered, the outlet valve should open but only if the temperature sensor indicates the molasses is not too cold. The valve should remain open until the lower sensor is uncovered or if the molasses cools off so that it will not flow properly.

The alarm logic is designed to alert operators if a failure is detected in a sensor. This can occur if

the upper level sensor indicates that it is covered but the lower sensor does not indicate that it is covered. Under this condition, an alarm should be indicated by an LED. The active level for the alarm should be LOW. Construct and test the logic for V_{OUT} and A.

Report

You will need to prepare a truth table (similar to Table 10–2) for each output you are designing. Comments are optional but can help clarify the reasons behind the logic for a given output. You should also prepare a Karnaugh map for each output

TABLE 10–2
Truth table for the output valve of the model-2 tank controller.

Inputs				Output	
L_H	L_L	T_C	V_{OUT}	V_{OUT}	Action
0	0	0	0	1	Close valve.
0	0	0	1	1	Leave valve closed.
0	0	1	0	1	Close valve.
0	0	1	1	1	Leave valve closed.
0	1	0	0	0	Valve is already open; leave open.
0	1	0	1	1	Valve is closed; leave closed.
0	1	1	0	1	Close valve; temp too cold.
0	1	1	1	1	Leave valve closed; temp too cold.
1	0	0	0	0	Sensor error; open valve.
1	0	0	1	0	Sensor error; open valve.
1	0	1	0	0	Sensor error; open valve.
1	0	1	1	0	Sensor error; open valve.
1	1	0	0	0	Tank full; leave valve open.
1	1	0	1	0	Tank full; open valve.
1	1	1	0	1	Tank full but too cold, close valve.
1	1	1	1	1	Tank full but too cold, keep valve closed.

that you design and show the simplified logic expression. A single blank Karnaugh map is shown in Figure 10–4 that has the same inputs as the truth table. Again, notice that one of the inputs is listed as V_{OUT} because of the feedback.

After reading the maps, draw a schematic of the circuit. Write a technical report summarizing it and your test results. Your instructor may have a required format, or you may use the one given in the Introduction to the Student.

For Further Investigation

Construct and test the logic for the inlet valve, V_{IN}. The valve should be open only when the outlet valve is closed and the upper level sensor is not covered. Describe the results of your test in your report.

Multisim Troubleshooting (Optional)

The companion website for this manual has Multisim 11 and Multisim 12 files. The simulated circuit is the model-1 outlet valve logic (Figure 10–2). It is

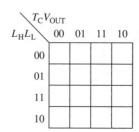

FIGURE 10–4
Blank Karnaugh map for mapping an output.

a good way to see the effect of the feedback in this application. Download the file named Exp-10nf and the worksheet Exp-10ws. Open the file named Exp-10nf. Complete the worksheet and attach it to the report.

This circuit is also simulated in Quartus II, which is discussed in Experiment 26. In this case, the simulation has waveforms to view.

Experiment 11
Adder and Magnitude Comparator

Objectives

After completing this experiment, you will be able to

□ Complete the design, build, and test a 4-bit binary to Excess-3 code converter.

□ Complete the design of a signed number adder with overflow detection.

Materials Needed

7483A 4-bit binary adder
7485 4-bit magnitude comparator
7404 hex inverter
Five LEDs
One 4-position DIP switch
Resistors: five 330 Ω, eight 1.0 kΩ

For Further Investigation:

Materials to be determined by student

Summary of Theory

This experiment introduces you to two important MSI circuits—the 4-bit adder and the 4-bit magnitude comparator. The TTL version of a 4-bit adder, with full look-ahead carry, is the 7483A. The TTL version of the 4-bit magnitude comparator is the 7485 which includes $A > B$, $A < B$, and $A = B$ outputs. It also has $A > B$, $A < B$, and $A = B$ inputs for the purpose of cascading comparators. Be

careful to note inputs and outputs when you are connecting a comparator.

One difference in the way adders and comparators are labeled needs to be clarified. A 4-bit adder contains a carry-in, which is considered the least significant bit so it is given a *zero* subscript (C_0). The next least significant bits are contained in the two 4-bit numbers to be added, so they are identified with a *one* subscript (A_1, B_1). On a comparator, there is no carry-in, so the least significant bits are labeled with a *zero* subscript (A_0, B_0).

In this experiment, an adder and comparator are used to convert a 4-bit binary code to Excess-3 code. The approach is somewhat unorthodox but shows how adders and magnitude comparators work. Although the technique could be applied to larger binary numbers, better methods are available. To familiarize you with the experiment, a similar circuit is described in the following example.

Example: A 4-Bit Binary to BCD Converter

Recall that a 4-bit binary number between 0000 and 1001 (decimal 9) is the same as the BCD number. If we add zero to these binary numbers, the result is unchanged and still represents BCD. Binary numbers from 1010 (ten) to 1111 (fifteen) can be converted to BCD by simply adding 0110 (six) to them.

The circuit that accomplishes this is illustrated in Figure 11–1. Notice that the *B* input of the comparator is connected to 1001. If the binary number

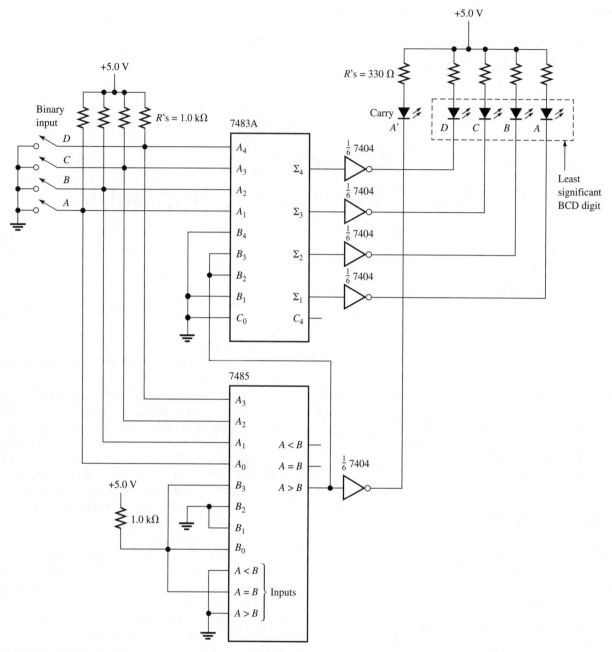

FIGURE 11–1
Binary to BCD converter.

on the A input of the comparator is greater than 1001, the $A > B$ output is asserted. This action causes the adder to add 0110 to the binary input. Notice how the $A > B$ output of the comparator is connected to the B side of the adder. Only bits B_2 and B_3 will be HIGH when $A > B$ is asserted; otherwise, all bits are LOW. This causes the adder to either add 0000 or 0110 to the binary input. The situation is summarized in Table 11–1. Notice in the table that bits B_2 and B_3 have the same "sense" as the $A > B$ output—they are 0 when $A > B$ is 0 and 1

when $A > B$ is 1. Bits B_4 and B_1 are always 0, hence these inputs are connected to ground.

Procedure

1. Figure 11–2 (in the report) shows a partially completed schematic of a binary to Excess-3 code conversion circuit. It uses the basic idea described by the example in the Summary of Theory except it needs to add either 0011 or 1001 to the binary input number. The adder must add 0011

TABLE 11–1

Binary to BCD.

Comparator $A > B$ Output	Adder Input				Comment
	B_4	B_3	B_2	B_1	
0	0	0	0	0	input is less than ten, add 0000
1	0	1	1	0	input is greater than nine, add 0110

(decimal 3) to the binary input number if it is between 0000 and 1001 but add 1001 (nine) to the number if it is greater than nine in order to convert the 4-bit binary number to Excess-3. The problem is summarized in Table 11–2. Decide how to connect the open inputs on the 7483A, and complete the schematic.

2. From your schematic, build the circuit. Test all possible inputs as listed on truth Table 11–4 (in the report). The outputs can be read directly from the LEDs. Assume an LED that is ON represents a logic 1 and an LED that is OFF represents a logic 0.

For Further Investigation

Overflow Detection

Fixed-point signed numbers are stored in most computers in the manner illustrated in Table 11–3. Positive numbers are stored in true form and negative numbers are stored in 2's complement form. If two numbers with the same sign are added, the answer can be too large to be represented with the number of bits available. This condition, called *overflow*, occurs when an addition operation causes a carry into the sign bit position. As a result, the sign bit will be in error, a condition easy to detect.

When two numbers with the opposite sign are added, overflow cannot occur, so the sign bit will always be correct. Figure 11–3 illustrates overflow for 4-bit numbers.

In this part of the experiment we will step through the design of a 4-bit adder for signed numbers that detects the presence of an overflow error and lights an LED when overflow occurs. We can start with the 7483A adder and a 7404 hex inverter as shown in Figure 11–4 (in the report).

1. Consider the problem of detecting an overflow error. We need consider only the sign bit for each number to be added and the sign bit for the answer. Complete truth Table 11–5 for all possible combinations of the sign bit, showing a 1 whenever an overflow error occurs.

2. Complete the Karnaugh map of the output (shown in the report as Figure 11–5) to see whether minimization is possible.

3. Write the Boolean expression for detection of an overflow error in your report.

4. Note that the signals going into the box in Figure 11–4 are A_4, B_4, and $\overline{\Sigma}_4$. If you apply DeMorgan's theorem to one term of your Boolean expression, you can draw a circuit that uses only these inputs. Draw the circuit in the box. If directed by your instructor, build and test your circuit.

TABLE 11–2

Binary to Excess-3.

Comparator $A > B$ Output	Adder Input				Comment
	B_4	B_3	B_2	B_1	
0	0	0	1	1	input is less than ten, add 0011
1	1	0	0	1	input is greater than nine, add 1001

TABLE 11–3
Representation of 4-bit signed numbers.

Base Ten Number	Computer Representation	
+7	0111	
+6	0110	
+5	0101	
+4	0100	Numbers in
+3	0011	true form
+2	0010	
+1	0001	
0	0000	
−1	1111	
−2	1110	
−3	1101	
−4	1100	Numbers in 2's
−5	1011	complement
−6	1010	form
−7	1001	
−8	1000	

Sign bit

Numbers with opposite signs: Overflow into sign position cannot occur.

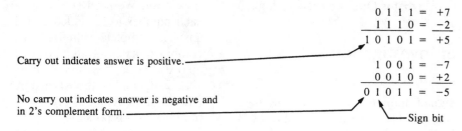

Carry out indicates answer is positive.

No carry out indicates answer is negative and in 2's complement form.

Numbers with the same sign: Overflow into sign position can occur.

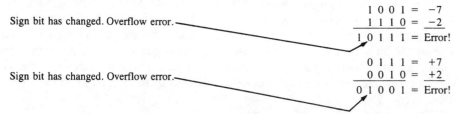

Sign bit has changed. Overflow error.

Sign bit has changed. Overflow error.

FIGURE 11–3
Overflow that can occur with 4-bit signed numbers. The concept shown here is applied to large binary numbers.

Report for Experiment 11

Name: _____ Date: _____ Class: _____

Objectives:

☐ Complete the design, build, and test a 4-bit binary to Excess-3 code converter.
☐ Complete the design of a signed number adder with overflow detection.

Data and Observations:

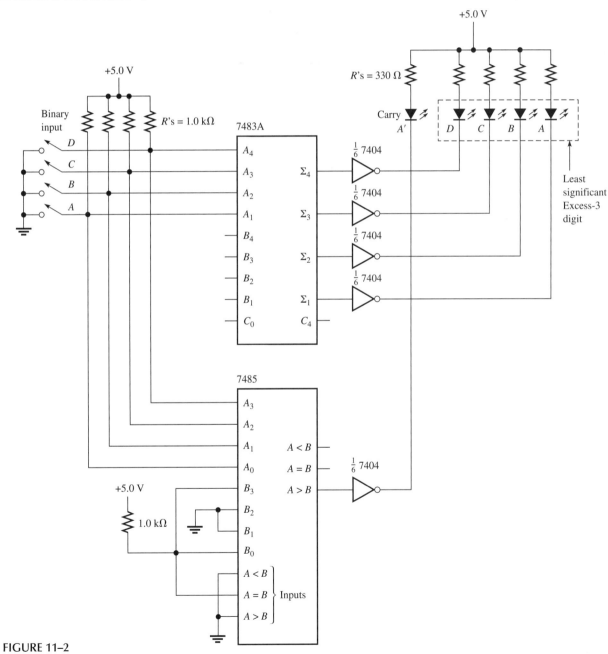

FIGURE 11–2
Binary to Excess-3 converter.

TABLE 11–4
Truth table for Figure 11–2.

Inputs (Binary)				Outputs Excess-3				
D	C	B	A	A'	D	C	B	A
0	0	0	0					
0	0	0	1					
0	0	1	0					
0	0	1	1					
0	1	0	0					
0	1	0	1					
0	1	1	0					
0	1	1	1					
1	0	0	0					
1	0	0	1					
1	0	1	0					
1	0	1	1					
1	1	0	0					
1	1	0	1					
1	1	1	0					
1	1	1	1					

Results and Conclusion:

Further Investigation Results:

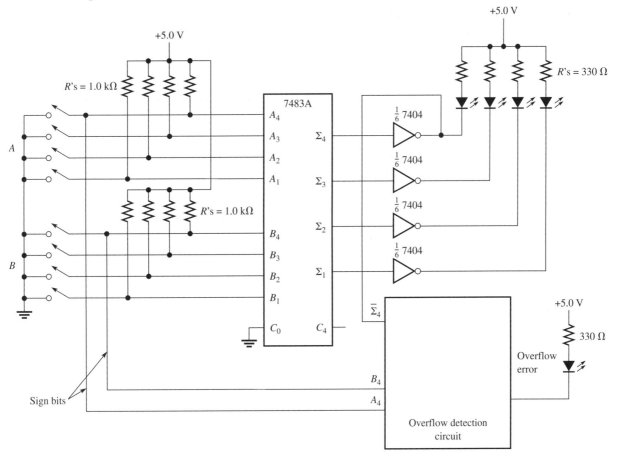

FIGURE 11–4
Signed number adder.

TABLE 11–5
Truth table for overflow error.

Sign Bits			Error
A_4	B_4	Σ_4	X
0	0	0	
0	0	1	
0	1	0	
0	1	1	
1	0	0	
1	0	1	
1	1	0	
1	1	1	

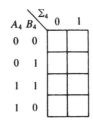

FIGURE 11–5
Karnaugh map for overflow error.

Boolean expression for overflow error:

$X =$ _____

Evaluation and Review Questions

1. Assume the circuit of Figure 11–1 had an open output on the $A > B$ of the 7485.
    a. What effect does this have on the A' output?

    b. What voltage level would you expect to measure on the B_2 and B_3 inputs on the 7483A?

2. In the space below, show how two 7483A adders could be cascaded to add two 8-bit numbers.

3. What is the function of the C_0 input on the 7483A adder?

4. The circuit of Figure 11–6 should turn on the LED whenever the input number is less than 8 or greater than 12. Complete the design by showing where each of the remaining inputs, including the cascading inputs, is connected.

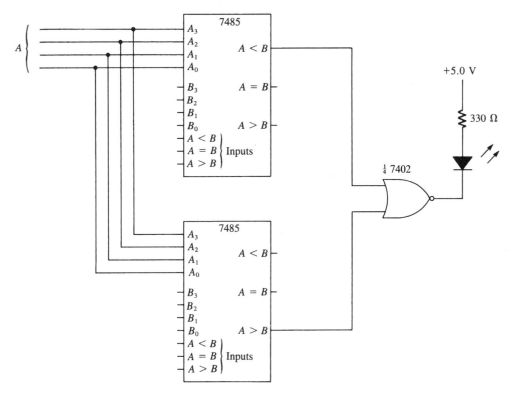

FIGURE 11–6

5. The circuit in Figure 11–4 is designed to add two 4-bit signed numbers. Recall that negative numbers are stored in 2's complement form. To change the circuit into a subtracting circuit $(A - B)$ you need to 2's complement the B (subtrahend) inputs. Inverting the B inputs forms the 1's complement. What else must be done in order to cause the circuit to form the 2's complement of the B input?

6. Figure 11–7 (on the following page) shows a simple six-position voting machine module using CMOS logic. What change could you make to the module to allow a seventh position? (*Hint:* no additional ICs are needed!)

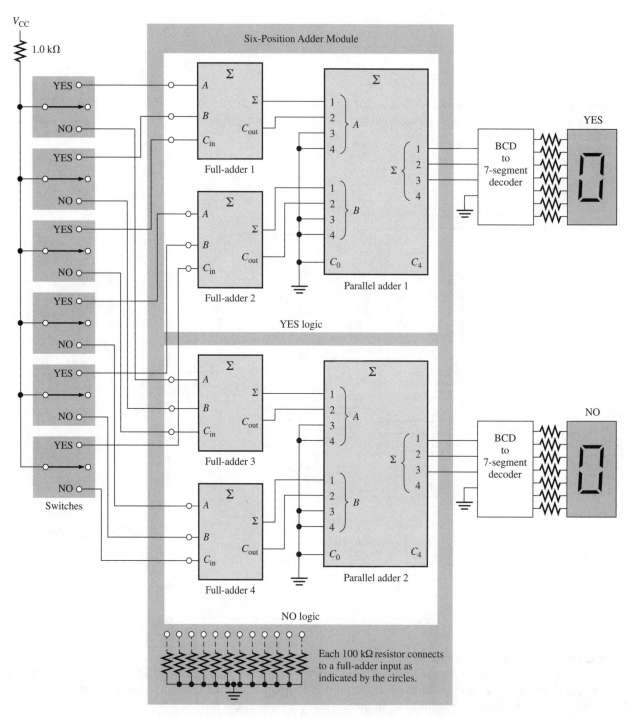

FIGURE 11–7

90

Experiment 12
Combinational Logic Using Multiplexers

Objectives

After completing this experiment, you will be able to

☐ Use a multiplexer to construct a comparator and a parity generator and test the circuit.

☐ Use an *N*-input multiplexer to implement a truth table containing 2*N* inputs.

☐ Troubleshoot a simulated failure in a test circuit.

Materials Needed

74151A data selector/multiplexer
7404 hex inverter
One LED
Resistors: one 330 Ω, four 1.0 kΩ

Summary of Theory

The *multiplexer* or *data selector* connects any one of several inputs to a single output. The opposite function, in which a single input is directed to one of several outputs, is called a *demultiplexer* or a *decoder*. These definitions are illustrated in Figure 12–1. Control is determined by additional logic signals called the *select* (or *address*) inputs.

Multiplexers (MUXs) and demultiplexers (DMUXs) have many applications in digital logic. One useful application for MUXs is implementation of combinational logic functions directly from the truth table. For example, an overflow error detection circuit is described by the truth table shown in

Figure 12–2(a). Overflow is an error that occurs when the addition of signed numbers produces an answer that is too large for the register to which it is assigned. The truth table indicates that output X should be logic 1 (indicating overflow) if the inputs (labeled A_4, B_4, Σ_4) are 0, 0, 1, or 1, 1, 0, which happen when there is an overflow error. Notice that each of the lines on the truth table corresponds to one of the 8 inputs on the MUX. On the MUX, the 001 and 110 lines are lines D_1 and D_6. The data itself controls which line is active by connecting it to the Select inputs. The D_1 and D_6 lines are tied HIGH, so when they are selected, the output of the MUX will also be HIGH. All other data inputs on the MUX are tied to a LOW. When one of the other inputs is selected, a LOW is routed to the output, thus implementing the truth table. This idea is shown conceptually in the diagram in Figure 12–2(b).

Actually, an 8-input MUX is not required to implement the overflow detection logic. Any *N*-input MUX can generate the output function for 2*N* inputs. To illustrate, we reorganize the truth table in pairs, as shown in Figure 12–3(a). The inputs labeled A_4 and B_4 are used to select a data line. Connected to that line can be a logic 0, 1, Σ_4, or $\overline{\Sigma}_4$. For example, from the truth table, if $A_4 = 0$ and $B_4 = 1$, the D_1 input is selected. Since both outputs are the same (in this case a logic 0), then D_1 is connected to a logic 0. If the outputs were different, such as in the first and fourth rows, then the third input variable, Σ_4, would be compared with the output. Either the

FIGURE 12–1

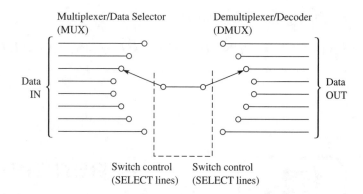

true (or NOT) form of that variable then would be selected. The results are shown conceptually in Figure 12–3(b), which is equivalent to but simpler than the idea illustrated in Figure 12–2(b).

In this experiment you will use an 8:1 MUX to implement a 4-input truth table (with 16 combinations). First you will develop the circuit to implement a special comparator. In the For Further Investigation section the circuit is modified to make a parity generator for a 4-bit code. *Parity* is an extra bit attached to a code to check that the code has been received correctly. *Odd parity* means that the number of 1's in the code, including the parity bit, is an odd number. Odd or even parity can be generated with exclusive-OR gates, and parity generators are available in IC form. However, the implement-

ing of an arbitrary truth table using MUXs is the important concept.

Procedure

Special 2-Bit Comparator

1. Assume you needed to compare two 2-bit numbers called A and B to find whether A is equal to or greater than B. You could use a comparator and OR the $A > B$ and $A = B$ outputs. Another technique is to use an 8:1 MUX with the method shown in the Summary of Theory section. The partially completed truth table for the comparator is shown as Table 12–1 in the report. The inputs are A_2, A_1 and B_2, B_1, representing the two numbers to be compared. Notice that the A_2, A_1, and B_2 bits are connected to

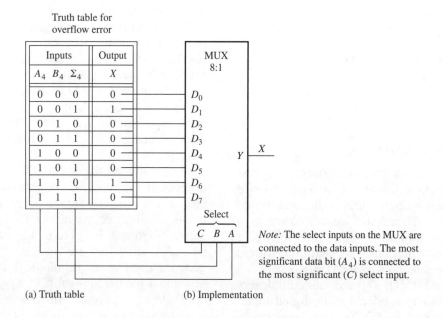

(a) Truth table (b) Implementation

Note: The select inputs on the MUX are connected to the data inputs. The most significant data bit (A_4) is connected to the most significant (C) select input.

FIGURE 12–2

92

Truth table for overflow error

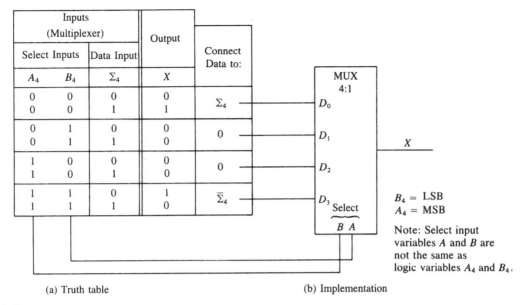

Inputs (Multiplexer)			Output		
Select Inputs		Data Input		Connect Data to:	
A_4	B_4	Σ_4	X		
0	0	0	0	Σ_4	
0	0	1	1		
0	1	0	0	0	
0	1	1	0		
1	0	0	0	0	
1	0	1	0		
1	1	0	1	$\overline{\Sigma_4}$	
1	1	1	0		

(a) Truth table

(b) Implementation

MUX 4:1
D_0
D_1
D_2
D_3
Select
B A
X
$B_4 = $ LSB
$A_4 = $ MSB
Note: Select input variables A and B are not the same as logic variables A_4 and B_4.

FIGURE 12–3

the SELECT inputs of the MUX. The B_1 bit is available to be connected as needed. It is therefore listed in a separate column of the truth table. Determine the logic in which the output represents $A > B$ and complete the X column of truth Table 12–1. The first two entries are completed as an example.

2. Look at the output X, in groups of two. The first pair of entries in X is the complement of the corresponding entries in B_1; therefore, the data should be connected to $\overline{B_1}$, as shown in the first line. Complete Table 12–1 by filling in the last column with either 0, 1, B_1, or $\overline{B_1}$.

3. Using the data from Table 12–1, complete the circuit drawing shown as Figure 12–4 in the report. The X output on the truth table is equivalent to the Y output on the 74151A. From the manufacturer's data sheet,* determine how to connect the STROBE input (labeled $\overline{G}$). Construct the circuit and test its operation by checking every possible input. Demonstrate your working circuit to your instructor.

For Further Investigation

Parity Generator Using a Multiplexer

1. The technique to implement an arbitrary function can also generate either odd or even parity. The MUX can generate *both* odd and even parity at the same time because there are two complementary outputs. One interesting aspect of the parity generator circuit is that any of the four inputs can turn the

LED on or off in a manner similar to the way in which 3-way switches can turn a light on or off from more than one location. The truth table is shown in the report as Table 12–2. Four of the bits (A_3 through A_0) represent the information, and the fifth bit (X), which is the output, represents the parity bit (this will be taken from the Y output of the 74151A). The requirement is for a circuit that generates both odd and even parity; however, the truth table will be set up for even parity. Even parity means that the sum of the 5 bits, *including the output parity bit,* must be equal to an even number. Complete truth Table 12–2 to reflect this requirement. The first line has been completed as an example. The even parity bit is taken from the Y output of the 74151A. The $\overline{W}$ output will be LOW when Y is HIGH and lights the LED.

2. Using the truth table completed in Step 1, complete the schematic for the even parity generator that is started in Figure 12–5 of the report. Change your original circuit into the parity circuit and test its operation.

Multisim Troubleshooting (Optional)

The companion website for this manual has Multisim 11 and Multisim 12 files. Download the file named Exp-12nf and the worksheet Exp-12ws. Open the file named Exp-12nf. The overflow error problem discussed in the Summary of Theory and illustrated in Figure 12–3 is given as Exp-12nf using the 74LS153 IC. Complete the worksheet and attach it to the report.

*See Appendix A.

Report for Experiment 12

Name: _____ Date: _____ Class: _____

Objectives:

☐ Use a multiplexer to construct a comparator and a parity generator and test the circuit.
☐ Use an N-input multiplexer to implement a truth table containing $2N$ inputs.
☐ Troubleshoot a simulated failure in a test circuit.

Data and Observations:

TABLE 12–1
Truth table for 2-bit comparator, $A \geq B$.

Inputs				Output	Connect Data to:
A_2	A_1	B_2	B_1	X	
0	0	0	0	1	$\overline{B_1}$
0	0	0	1	0	
0	0	1	0		
0	0	1	1		
0	1	0	0		
0	1	0	1		
0	1	1	0		
0	1	1	1		
1	0	0	0		
1	0	0	1		
1	0	1	0		
1	0	1	1		
1	1	0	0		
1	1	0	1		
1	1	1	0		
1	1	1	1		

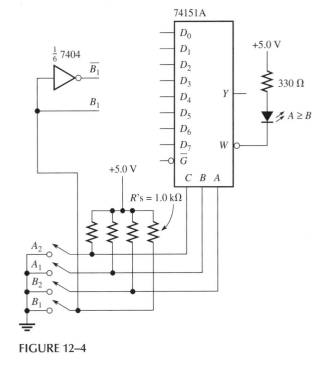

FIGURE 12–4

95

Results and Conclusion:

Further Investigation Results:

TABLE 12–2
Truth table for even parity generator.

Inputs				Output	Connect Data to:
A_3	A_2	A_1	A_0	X	
0	0	0	0	0	A_0
0	0	0	1	1	
0	0	1	0		
0	0	1	1		
0	1	0	0		
0	1	0	1		
0	1	1	0		
0	1	1	1		
1	0	0	0		
1	0	0	1		
1	0	1	0		
1	0	1	1		
1	1	0	0		
1	1	0	1		
1	1	1	0		
1	1	1	1s		

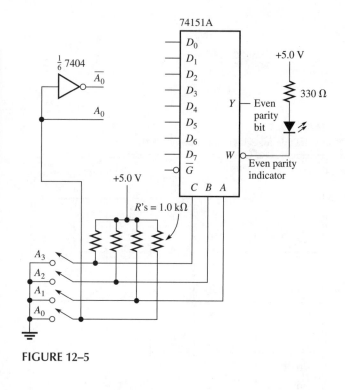

FIGURE 12–5

Evaluation and Review Questions

1. Design a BCD invalid code detector using a 74151A. Show the connections for your design on Figure 12–6.

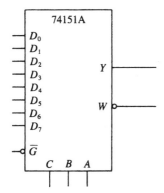

FIGURE 12–6

2. Can you reverse the procedure of this experiment? That is, given the circuit, can you find the Boolean expression? The circuit shown in Figure 12–7 uses a 4:1 MUX. The inputs are called A_2, A_1, and A_0. The first term is obtained by observing that when both select lines are LOW, A_2 is routed to the output; therefore the first minterm is written $A_2 \overline{A_1} \overline{A_0}$. Using this as an example, find the remaining minterms.

$X = A_2 \overline{A_1} \overline{A_0} +$ _____

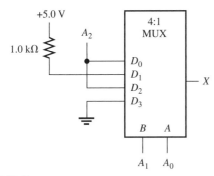

FIGURE 12–7

3. Assume the circuit shown in Figure 12–4 had the correct output for the first half of the truth table but had some incorrect outputs for the second half of the truth table. You decide to change ICs (not necessarily the best choice!) but the problem persists. What is the most likely cause of the problem? How would you test the circuit for your suspected problem?

4. Assume the circuit in Figure 12–4 had a short to ground on the output of the inverter. What effect would this have on the output logic? What procedure would you use to find the problem?

5. Assume that the input to the 7404 in Figure 12–4 was open, making the output, $\overline{B_1}$, a constant LOW. Which lines on the truth table would give incorrect readings on the output?

6. How can both odd and even parity be obtained from the circuit in Figure 12–5?

Experiment 13
Combinational Logic Using DeMultiplexers

Objectives

After completing this experiment, you will be able to

☐ Complete the design of a multiple output combinational logic circuit using a demultiplexer.

☐ Use an oscilloscope to develop a timing diagram for a counter-decoder circuit.

Materials Needed

7408 or 74LS08 quad AND gate
7474 dual D flip-flop
74LS139A decoder/demultiplexer
One 4-position DIP switch
LEDs: two red, two yellow, two green
Resistors: six 330 Ω, two 1.0 kΩ

For Further Investigation:
  7400 quad NAND gate

Summary of Theory

This experiment familiarizes you with the demultiplexer function, introduced in Chapter 5 of the text. The system introduced here is described in more detail in Section 6-1 of the text.

The demultiplexer (DMUX) can serve as a decoder or a data router. In this experiment, we will focus on the decoding function. A decoder takes binary information from one or more input lines and generates a unique output for each input combination. You are already familiar with the 7447A IC, which performs the decoding function. It converts a 4-bit binary input number to a unique code that is used to drive a 7-segment display. A DMUX can serve as a decoder by providing a unique output for every combination of input variables. The input variables are applied to the decoder's SELECT lines.

For most DMUXs, the selected output is LOW, whereas all others are HIGH. To implement a truth table that has a *single* output variable with a decoder is not very efficient and is rarely done; however, a method for doing this is shown conceptually in Figure 13–1. In this case, each line on the output represents one row on the truth table. If the decoder has active-HIGH outputs, the output lines on the truth table with a 1 are ORed together, as illustrated in Figure 13–1. The output of the OR gate represents the output function. If the outputs of the decoder are active-LOW, the output lines with a 1 on the truth table are connected with a NAND gate. This is shown in Figure 13–2.

A DMUX is superior for implementing combinational logic when there are several outputs for the same set of input variables. As you saw, each output line of the demultiplexer represents a line on the truth table. For active-HIGH decoder outputs, OR gates are used, but a separate OR gate is required for each output function. Each OR gate output represents a different output function. In the case of active-LOW decoder outputs, the OR gates are replaced by NAND gates.

FIGURE 13–1

Implementing a combinational logic function with an active HIGH DMUX.

Truth table for overflow error

Inputs			Output
A_4	B_4	Σ_4	X
0	0	0	0
0	0	1	1
0	1	0	0
0	1	1	0
1	0	0	0
1	0	1	0
1	1	0	1
1	1	1	0

(a) Truth table

(b) Implementation

The problem presented in this experiment is the output logic for a traffic signal controller.* A brief synopsis of the problem is as follows:

A digital controller is required to control traffic at an intersection of a busy main street and an occasionally used side street. The main street is to have a green light for a minimum of 25 seconds or as long as there is no vehicle on the side street. The side street is to have a green light until there is no vehicle on the side street or for a maximum of 25 seconds. There is to be a 4-second caution light (yellow) between changes from green to red on both the main

*This traffic light system is described in Section 6-1 of the text.

street and the side street. These requirements are illustrated in the pictorial diagram in Figure 13–3. A block diagram of the system, showing the essential details, is given in Figure 13–4.

We will focus on combinational logic in this experiment. The key elements shown in Figure 13–4 can be separated into a state decoder, a light output logic block, and trigger logic as shown in Figure 13–5. The state decoder has two inputs (2-bit Gray code) and must have an output for each of the four states. The 74LS139A is a dual 2-line to 4-line decoder and will do the job nicely, so it is selected.

The light output logic takes the four active-LOW states from the decoder and must produce six

FIGURE 13–2

Truth table for overflow error

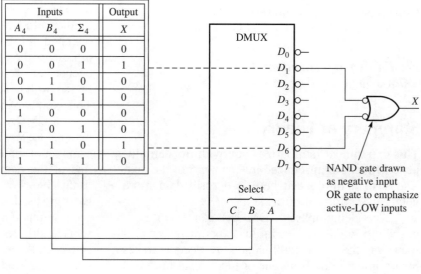

Inputs			Output
A_4	B_4	Σ_4	X
0	0	0	0
0	0	1	1
0	1	0	0
0	1	1	0
1	0	0	0
1	0	1	0
1	1	0	1
1	1	1	0

(a) Truth table

(b) Implementation

100

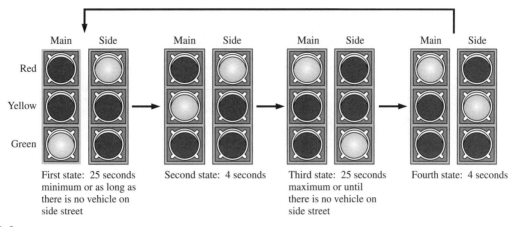

FIGURE 13–3
Requirements for the traffic light sequence.

First state: 25 seconds minimum or as long as there is no vehicle on side street

Second state: 4 seconds

Third state: 25 seconds maximum or until there is no vehicle on side street

Fourth state: 4 seconds

outputs for activating the traffic lights. A truth table for the decoding and output logic is given in Table 13–1. The truth table is organized in Gray code, which is used by the sequential logic to step through the states. Unlike the text, the state outputs ($\overline{S_1}$ to $\overline{S_4}$) are active-LOW (0) and the output logic must be active-LOW (0) to drive LEDs that simulate the traffic lights. Aside from this difference, the circuit works the same.

Procedure

Traffic Signal Decoder

The circuit represents the state decoder for four states and the light output logic of a traffic signal controller system as described in the Summary of Theory section. There are six outputs representing a red, yellow, and green traffic light on a main and side street. The outputs are shown in the desired sequence for the lights on the truth Table 13–1. A logic "0" is used to turn on an LED. For example, state 00 (the first row of the truth table) will have a green light ON for the main street and a red light ON for the side street.

1. A partially completed schematic is shown in Figure 13–6 in the report. The 74LS139A is the state decoder and the AND gates, drawn as negative NOR gates, form the output logic. Refer to the truth table and complete the schematic. You need to decide what to do with the $1\overline{G}$ input and you should draw switches with pull-up resistors to each of the

FIGURE 13–4
System block diagram showing essential elements.

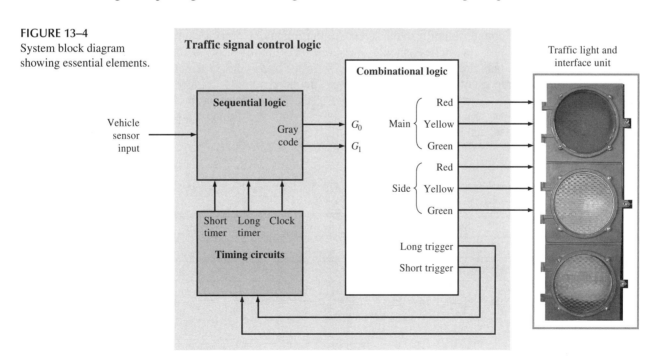

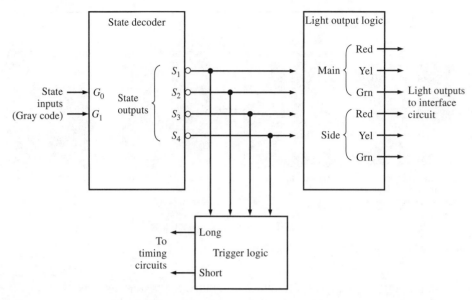

FIGURE 13–5
Block diagram of the state decoding and light output logic.

decoder's select inputs. The $1Y_0$ output of the decoder (state 00) has been connected to the green LED on the main street and the red LED on the side street as an example.

2. Construct the circuit and test every combination according to the truth table. You should be able to observe the correct traffic light sequence if you open and close the switches in the same sequence as listed in Table 13–1.

3. Although you have not studied counters yet, the Gray-code counter shown in Figure 13–7 is simple to build and is the basis of the sequential logic in the traffic signal control logic. Construct the counter and connect the counter outputs to the select inputs of the 74LS139A (switches should be removed). The select inputs are labeled $1B$ and $1A$ on the 74LS139A and represent the state inputs (G_1 and G_0) in Figure 13–5.

4. Set the pulse generator for a 1 Hz TTL level signal and observe the sequence of the lights. You

should see that the sequence matches the expected sequence for a traffic light, but the lights are not controlled by any inputs such as a vehicle sensor or timer. These improvements will be added in Experiment 24.

5. Observe the relative timing of the input and output signals from the decoder using an oscilloscope. Speed up the pulse generator to 10 kHz. Connect channel 1 of your oscilloscope to the $1B$ select input of the decoder, which is the MSB and changes slower than $1A$. Trigger the scope from channel 1. (You should always trigger from a slower signal to establish timing information and should not change the trigger channel for the timing measurements.) Connect channel 2 to $1A$ and observe the time relationship of the SELECT signals. Then move *only* the channel 2 probe to each of the decoder outputs (labeled $1Y_0$, $1Y_1$, and $1Y_3$ on the IC). These represent the state outputs $\overline{S_0}$, $\overline{S_1}$, $\overline{S_2}$, and $\overline{S_3}$. Plot the timing diagram in Figure 13–8 in the report.

TABLE 13–1
Truth table for the combinational logic. State outputs are active-LOW and light outputs are active-LOW. $\overline{MR}$ = main red; $\overline{MY}$ = main yellow; etc.

State Code		State Outputs				Light Outputs					
G_1	G_0	$\overline{S_1}$	$\overline{S_2}$	$\overline{S_3}$	$\overline{S_4}$	$\overline{MR}$	$\overline{MY}$	$\overline{MG}$	$\overline{SR}$	$\overline{SY}$	$\overline{SG}$
0	0	0	1	1	1	1	1	0	0	1	1
0	1	1	0	1	1	1	0	1	0	1	1
1	1	1	1	0	1	0	1	1	1	1	0
1	0	1	1	1	0	0	1	1	1	0	1

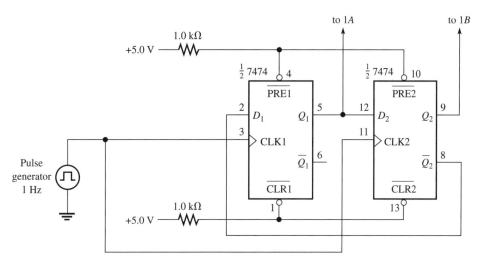

FIGURE 13–7
Gray-code counter for sequencing the traffic signal decoder.

For Further Investigation

An interesting application of both multiplexer (MUX) and DMUX is in time division multiplexing. Time division multiplexing is often done with displays. In this case, the DMUX is used to turn on only one 7-segment display at a time.

Time-division multiplexing is also applied to some data transmission systems. In this application, data is sent to the enable input of the DMUX and routed to the appropriate location by the DMUX. Naturally, the data and SELECT inputs have to be carefully synchronized.

A few changes to the circuit for this experiment will produce a similar idea. The simulated data is applied to the enable ($\overline{G}$) input. The particular output is addressed by the SELECT inputs, and synchronization is achieved by using the counter from Step 3 of the experiment to provide both the data and the address. The modified circuit, including the counter, is shown in Figure 13–9 in the report. The inputs to the NAND gate are not shown. Start by connecting them to the Q outputs of the counter.

Connect the circuit and observe the results. Notice the visual appearance of the LEDs. Compare the visual appearance with the waveform viewed on the oscilloscope. Try moving the inputs of the NAND gate to the $\overline{Q}$ outputs of the counter. What happens? Connect the NAND input to other combinations of the counter output. Summarize your findings in the report.

Report for Experiment 13

Name: _____ Date: _____ Class: _____

Objectives:

☐ Complete the design of a multiple output combinational logic circuit using a demultiplexer.
☐ Use an oscilloscope to develop a timing diagram for a counter-decoder circuit.

Data and Observations:

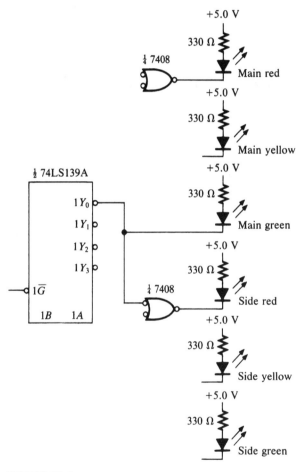

FIGURE 13–6
Traffic light output logic.

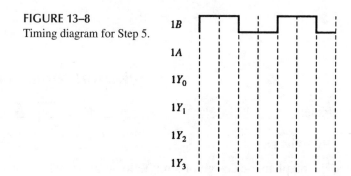

FIGURE 13–8
Timing diagram for Step 5.

Results and Conclusion:

Further Investigation Results:

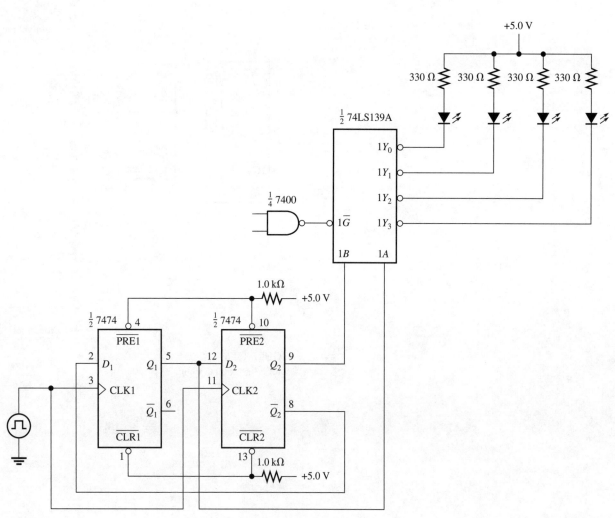

FIGURE 13–9

Evaluation and Review Questions

1. Assume you needed an 8-bit decoder, but all that is available is a dual 74LS139A. Show how you could use it, along with an inverter, to form an 8-bit decoder. (*Hint:* Consider using the enable inputs.)

2. Why were the AND gates in Figure 13–6 drawn as negative-NOR gates?

3. What is the advantage of Gray code for the state sequence?

4. For the circuit in Figure 13–6, what symptom would you observe if:
   a. The 1B select input were open?

   b. The 1B select input were shorted to ground?

   c. The enable ($\overline{G}$) input were open?

5. In Step 3, you added a Gray-code counter to the circuit. How would the circuit be affected if instead of a Gray-code counter, you had used a binary counter with the sequence 00-01-10-11?

6. Figure 13–10 shows a 74LS139 decoder and an XOR gate. For this circuit, under what conditions will be LED be ON? (*Hint:* For the 8 combinations of input switches, five will light the LED.)

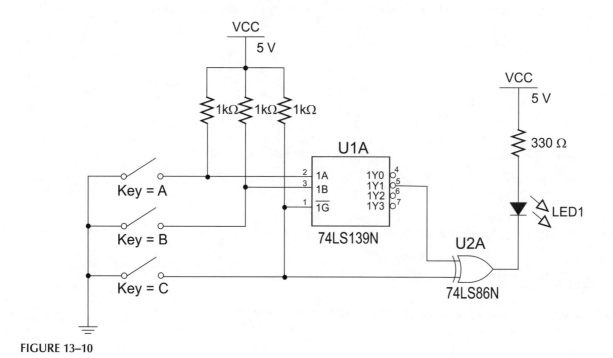

FIGURE 13–10

Experiment 14
The D Latch and D Flip-Flop

Objectives

After completing this experiment, you will be able to

☐ Demonstrate how a latch can debounce an SPDT switch.

☐ Construct and test a gated D latch from four NAND gates and an inverter.

☐ Test a D flip-flop and investigate several application circuits for both the latch and the flip-flop.

Materials Needed

Red LED
Green LED
7486 quad XOR gate
7400 quad NAND gate
7404 hex inverter
7474 dual D flip-flop
Resistors: two 330 Ω, two 1.0 kΩ

Summary of Theory

As you have seen, *combinational* logic circuits are circuits in which the outputs are determined fully by the inputs. *Sequential* logic circuits contain information about previous conditions. The difference is that sequential circuits contain *memory* and combinational circuits do not.

The basic memory unit is the *latch,* which uses feedback to lock onto and hold data. It can be constructed from two inverters, two NAND gates, or two NOR gates. The ability to remember previous conditions is easy to demonstrate with Boolean algebra. For example, Figure 14–1 shows an $\overline{\text{S}}$-$\overline{\text{R}}$ latch made from NAND gates. This circuit is widely used for switch debouncing and is available as an integrated circuit containing four latches (the 74LS279A).

A simple modification of the basic latch is the addition of steering gates and an inverter, as shown in Figure 14–2. This circuit is called a gated D (for Data) latch. An enable input allows data present on the D input to be transferred to the output when Enable is asserted. When the enable input is not asserted, the last level of Q and $\overline{Q}$ is latched. This circuit is available in integrated circuit form as the 7475A quad D latch. Although there are four latches in this IC, there are only two shared enable signals.

Design problems are often simplified by having all transitions in a system occur synchronously (at the same time) by using a common source of pulses to cause the change. This common pulse is called a *clock*. The output changes occur only on either the leading or the trailing edge of the clock pulse. Some ICs have inputs that directly set or reset the output any time they are asserted. These inputs are labeled *asynchronous* inputs because no clock pulse is required. The D-type flip-flop with positive edge-triggering and asynchronous inputs is the 7474. In this experiment, you will also test this IC.

FIGURE 14–1
$\overline{S}$-$\overline{R}$ latch.

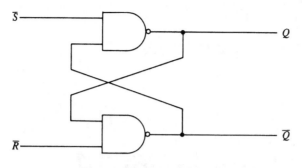

Equation for top NAND gate:

$$Q = \overline{\overline{S} \cdot \overline{Q}}$$

Applying DeMorgan's theorem:

$$Q = S + Q$$

Thus, Q appears on both sides of the equation.
If $\overline{S} = 1$, then $S = 0$ and $Q = 0 + Q$ (Q is previous state)
output is latched.

It is useful to review oscilloscope timing. If you are using an analog dual-trace oscilloscope, you should trigger the scope from the channel with the *slowest* of two waveforms that are being compared to be sure to see the correct time relationship. A digital scope will show it correctly for either trigger channel.

Procedure

$\overline{S}$-$\overline{R}$ Latch

1. Build the $\overline{S}$-$\overline{R}$ latch shown in Figure 14–3. You can use a wire to simulate the single-pole, double-throw (SPDT) switch. The LEDs will be used in this section as logic monitors. Because TTL logic is much better at sinking current than at sourcing current, the LEDs are arranged to be ON when the output is LOW. To make the LEDs read the HIGH output when they are ON, we read them from the opposite output! This simple trick avoids the use of an inverter.

2. Leave the wire on the *A* terminal and note the logic shown on the LEDs. Now simulate a

bouncing switch by removing the *A* end of the wire. Do NOT touch *B* yet! Instead, reconnect the wire to *A* several times.

3. After touching *A* several times, touch *B*. Simulate the switch bouncing several times by removing and reconnecting *B*. (Switches never bounce back to the opposite terminal, so you should not touch *A*). Summarize your observations of the $\overline{S}$-$\overline{R}$ latch used as a switch debounce circuit in the report.

D Latch

4. Modify the basic $\overline{S}$-$\overline{R}$ latch into a D latch by adding the steering gates and the inverter shown in Figure 14–4. Connect the *D* input to a TTL level pulse generator set for 1 Hz. Connect the enable input to a HIGH (through a 1.0 kΩ resistor). Observe the output; then change the enable to a LOW.

5. Leave the enable LOW and place a momentary short to ground first on one output and then on the other. Summarize your observations of the gated D latch in the report.

FIGURE 14–2
Gated D latch.

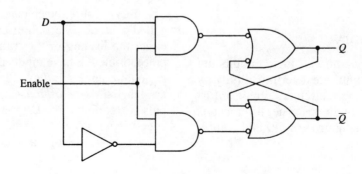

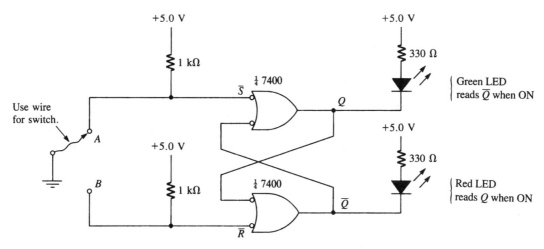

FIGURE 14–3
SPDT switch debounce. The NAND gates are drawn as negative-input OR gates to emphasize the active-LOW.

6. Now make the simple burglar alarm shown in Figure 14–5. The data input represents switches connected to windows and doors. The enable input is pulled HIGH when the system is activated or LOW for standby. To reset the system, put a momentary ground on the Q output as shown. Summarize your observations in the report.

The D Flip-Flop

7. The 7474 is a dual, positive edge–triggered D flip-flop containing two asynchronous inputs labeled $\overline{PRE}$ (preset) and $\overline{CLR}$ (clear). Construct the test circuit shown in Figure 14–6. Connect the clock through the delay circuit. The purpose of the delay is to allow setup time for the D input. Let's

look at this effect first. Observe both the delayed clock signal and the Q output signal on a two-channel oscilloscope. View the delayed clock signal on channel 1, and trigger the scope from channel 1. You should observe a dc level on the output (channel 2).

8. Now remove the clock delay by connecting the clock input directly to the pulse generator. The output dc level should have changed because there is insufficient setup time. Explain your observations in the report.

9. Reinstall the clock delay circuit and move the $\overline{PRE}$ input to a LOW and then a HIGH. Then put a LOW on the $\overline{CLR}$ input followed by a HIGH. Next repeat the process with the clock pulse disconnected. Determine if $\overline{PRE}$ and $\overline{CLR}$ are synchronous or asynchronous inputs.

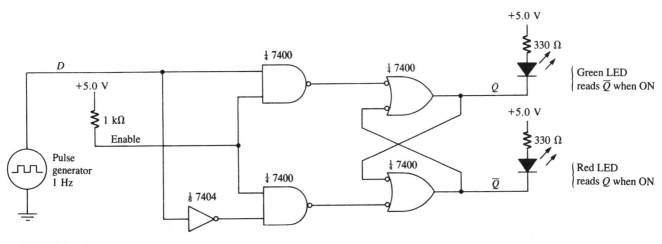

FIGURE 14–4
Gated D latch.

111

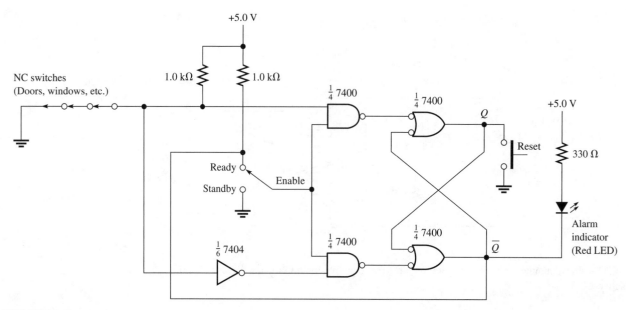

FIGURE 14–5
Simple burglar alarm.

10. Leave the clock delay circuit in place, but disconnect the D input. Attach a wire from $\overline{Q}$ to the D input. Observe the waveforms on a scope. Normally, for relative timing measurements, you should trigger the scope using the channel that has the *slowest* waveform as the trigger channel, as discussed in the Summary of Theory. Summarize, in the report, your observations of the D flip-flop. Discuss setup time, $\overline{PRE}$ and $\overline{CLR}$ inputs, and your timing observation from this step.

For Further Investigation

The circuit shown in Figure 14–7 is a practical application of a D flip-flop. It is a parity test circuit that takes serial data (bits arriving one at a time) and performs an exclusive-OR on the previous result (like a running total). The data are synchronous with the clock; that is, for every clock pulse a new data bit is tested. Construct the circuit and set the clock for 1 Hz. Move the data switch to either a HIGH or

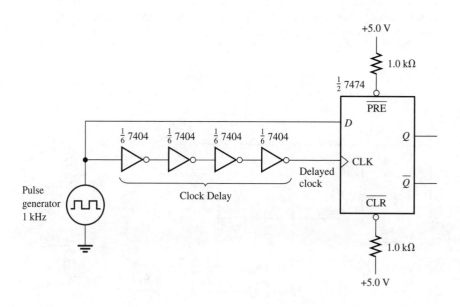

FIGURE 14–6

112

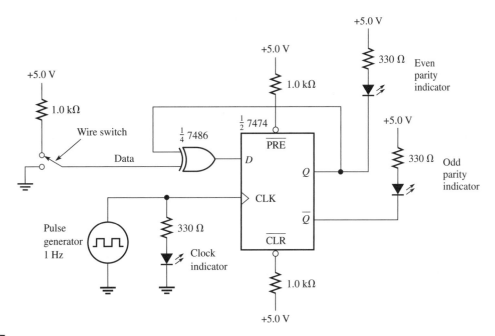

FIGURE 14–7

a LOW prior to the clock pulse, and observe the result. Describe your observations in the report. If a logic 1 is received, what happens to the parity? What happens when a logic 0 is received? Does the circuit have any advantage over the 9-bit parity generator/checker discussed in Section 5-11 of the text?

Report for Experiment 14

Name:_____ Date:_____ Class:_____

Objectives:

☐ Demonstrate how a latch can debounce an SPDT switch.
☐ Construct and test a gated D latch from four NAND gates and an inverter.
☐ Test a D flip-flop and investigate several application circuits for both the latch and the flip-flop.

Data and Observations:

Step 3. Observations for SPDT switch debounce circuit:

Step 5. Observations for D latch circuit:

Step 6. Observations for the simple burglar alarm:

Steps 7 and 8. Observations for setup time:

Step 10. Observations for the D flip-flop:

Results and Conclusion:

Further Investigation Results:

Evaluation and Review Questions

1. Explain why the switch debounce circuit in Figure 14–3 is used only for double-throw switches.

2. Could NOR gates be used for debouncing a switch? Explain.

3. Show how the burglar alarm of Step 6 could be constructed with one fourth of a 7475A D latch.

4. The burglar alarm could be constructed with two cross-coupled NOR gates. Complete the circuit so that it has the same functions as the alarm in Step 6 (normally closed switches, alarm, and reset).

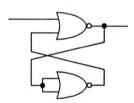

5. Assume that the burglar alarm in Figure 14–5 does not function correctly. The Enable switch is in the Ready position but the LED does not come on when one of the switches is open. Suggest at least 3 causes for this failure. Circle the reason you think is most likely.

6. The serial parity test circuit in Figure 14–7 uses a D flip-flop. Why wouldn't a D latch work just as well?

Experiment 15
The Fallen-Carton Detector

Objectives

After completing this experiment, you will be able to

☐ Design a circuit that detects the presence of a tipped-over carton for a food-processing application and rejects it before it reaches the carton-sealing machine.

☐ Decide on a troubleshooting procedure for testing the circuit if it fails.

☐ Write a formal laboratory report documenting your circuit and a simple test procedure.

Materials Needed

7474 D flip-flop
Two CdS photocells (Jameco 120299 or equivalent)
Other materials as determined by student

Summary of Theory

A D flip-flop can hold information temporarily, acting as a memory element capable of storing one bit of information. In this experiment, it is necessary to store information temporarily to use after the inputs have changed. The circuit can then take an action even though the originating event has passed.

The event can do the clocking action to assure that the flags are set each time an event occurs. The occurrence of an event (carton passing a detector) is asynchronous—not related to a clock signal. Since the event will do the clocking, it is necessary to use delay in the clock signal to assure that sufficient setup time is given to the D flip-flop.

Procedure

Design the circuit that implements the fallen-carton detector described in the problem statement. Test your circuit and write a report that describes the circuit you designed. The circuit will be in operation when you are not present to help fix it in case of trouble. Your write-up should include a simple troubleshooting guide to technicians so that they can identify a circuit failure using a test procedure that you devise.

Problem Statement: A food-processing company needs a circuit to detect a fallen carton on a conveyer belt and activate an air solenoid to blow the carton into a reject hopper. The circuit is needed because occasionally a carton will fall and cause a jam to occur at the carton-sealing machine. Mechanics have installed two photocells on the line, as illustrated in Figure 15–1. Notice that the two photocells are offset from each other. The top photocell is labeled *A* and the lower photocell is labeled *B*. An upright carton will cause photocell *A* to be covered first and then photocell *B* to be covered. A fallen carton will cause only photocell *B* to be covered.

You can assume that if both photocells sense the carton, it must be upright and should proceed to the sealing machine. However, if only photocell *B* is covered, a signal needs to be generated that will be sent to the air solenoid (an LED will simulate the

117

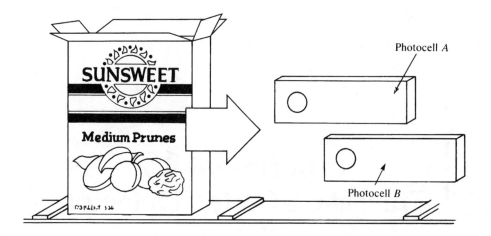

FIGURE 15–1

solenoid). The solenoid needs to be turned on until the next upright carton is sensed.

For this design problem, you can use the photocell circuit shown in Figure 15–2 to produce logic levels for your circuit. You will need to experiment with the value of R_1 to determine a value that gives TTL logic levels for your particular photocell and room lighting. The resistance of a photocell is lower as the light intensity increases. If the photocell is covered (high resistance), the output should be set for a logic HIGH; when it is uncovered it should be a logic LOW. If photocell A is covered, and then photocell B is covered, the circuit should *not* trip the solenoid (light the LED). On the other hand, if photocell B is covered and A remains uncovered, the circuit has detected a fallen carton, and the solenoid should be tripped (LED on). The LED should be turned on with a LOW signal. No more than two ICs can be spared for the design.

Report

Write a technical report summarizing your circuit and your results of testing it. Include a description of any tests you made to verify operation of the light sensors. Your instructor may have a required format,

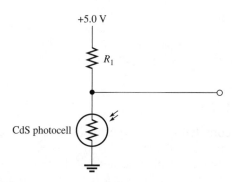

FIGURE 15–2
Simplified photocell circuit.

or you may use the one given in the Introduction to the Student.

For Further Investigation

A new improved Model 2 fallen-carton detector is needed. After a single fallen carton, the air solenoid remains on, annoying the line supervisor. To avoid this, the line supervisor has requested that the solenoid be turned off by a photocell that senses the carton going into the reject hopper. Also, a reset button needs to be added to reset the circuit when it is first turned on. No more ICs can be spared for this modification. Modify your design to meet these requirements.

CHAPTER 6: LATCHES, FLIP-FLOPS, AND TIMERS

Experiment 16
The J-K Flip-Flop

Objectives

After completing this experiment, you will be able to

☐ Test various configurations for a J-K flip-flop, including the asynchronous and synchronous inputs.

☐ Observe frequency division characteristics in the toggle mode.

☐ Measure the propagation delay of a J-K flip-flop.

Materials Needed

74LS76A dual J-K flip-flop
LEDs: one red, one green, one yellow
Resistors: three 390 Ω, four 1.0 kΩ
One 4-position DIP switch

Summary of Theory

The D flip-flop is an edge-triggered device that allows the output to change only during the active clock edge. The D flip-flop can only be set and reset, a limitation for some applications. Furthermore, it cannot be latched unless the clock pulses are removed—a condition that limits applications for this device. The S-R flip-flop could be latched but one combination of inputs is disallowed ($S = 1$, $R = 1$). A solution to these problems is the J-K flip-flop, which is basically a clocked S-R flip-flop with additional logic to replace the S-R invalid output

states with a new mode called *toggle*. Toggle causes the flip-flop to change to the state opposite to its present state. It is similar to the operation of an automatic garage door opener. If the button is pressed when the door is open, the door will close; otherwise, it will open.

The J-K flip-flop is the most versatile of the three basic flip-flops. All applications for flip-flops can be accomplished with either the D or the J-K flip-flop. The clocked S-R flip-flop is seldom used; it is used mostly as an internal component of integrated circuits. (See the 74LS165A shift register, for example.) The truth tables for all three flip-flops are compared in Figure 16–1. The inputs are labeled *J* (the set mode) and *K* (the reset mode) to avoid confusion with the S-R flip-flop.

A certain amount of time is required for the input of a logic gate to affect the output. This time, called the *propagation delay time,* depends on the logic family. In the For Further Investigation section, you will investigate the propagation delay for a J-K flip-flop.

The need to assure that input data do not affect the output until they are at the correct level led to the concept of edge-triggering, which is the preferred method to assure synchronous transitions. An older method that is sometimes used is *pulse-triggered* or *master-slave* flip-flops. In these flip-flops, the data are clocked into the master on the leading edge of the clock and into the slave on the trailing edge of the clock. It is imperative that the input data not change during the time the clock pulse is HIGH or

S-R flip-flop		
Inputs		Output
S	R	Q
0	0	Latched
0	1	0
1	0	1
1	1	Invalid

D flip-flop	
Input	Output
D	Q
No equivalent	
0	0
1	1
No equivalent	

J-K flip-flop		
Inputs		Output
J	K	Q
0	0	Latched
0	1	0
1	0	1
1	1	Toggle

FIGURE 16–1
Comparison of basic flip-flops.

the data in the master may be changed. J-K flip-flops are available as either edge- or pulse-triggered devices. The older 7476 is a dual pulse-triggered device; the 74LS76A is edge-triggered on the HIGH to LOW transition of the clock. Either type will work for this experiment.*

Procedure

The J-K Edge-Triggered Flip-Flop

1. Construct the circuit of Figure 16–2(a). The LEDs are logic monitors and are ON when their output is LOW. Select the inactive level (HIGH) for $\overline{PRE}$ and $\overline{CLR}$. Select the "set" mode by connecting J to a logic 1 and K to a logic 0. With the clock LOW (not active), test the effect of $\overline{PRE}$ and $\overline{CLR}$ by putting a logic 0 on each, one at a time. Are preset and clear inputs synchronous or asynchronous?

Put $\overline{CLR}$ on LOW; then pulse the clock by putting a HIGH, then a LOW, on the clock. Observe that the $\overline{CLR}$ input overrides the J input.

Determine what happens if both $\overline{PRE}$ and $\overline{CLR}$ are connected to a 0 at the same time. Summarize your observations from this step in the report.

2. Put both $\overline{PRE}$ and $\overline{CLR}$ on a logic 1. Connect a TTL level pulse generator set to 1 Hz on the clock input. Add an LED clock indicator to the pulse generator, as shown in Figure 16–2(b), so that you can observe the clock pulse and the outputs at the same time. Test all four combinations of J and K inputs while observing the LEDs.

Are data transferred to the output on the leading or the trailing edge of the clock?

Observe that the output frequency is not the same as the clock frequency in the toggle mode. Also note that the output duty cycle in the toggle mode is not the same as the clock duty cycle. This is a good way to obtain a 50% duty cycle pulse.

Summarize your observations in the report. Include a discussion of the truth table for the J-K flip-flop.

3. Look at the circuit shown in Figure 16–3. From your knowledge of the truth table, predict what it will do; then test your prediction by building it. Summarize your observations.

4. An application of the toggle mode is found in certain counters. Cascaded flip-flops can be used to perform frequency division in a circuit called a *ripple counter.*** Figure 16–4 illustrates a ripple counter using the two flip-flops in the 74LS76. Connect the circuit and sketch the Q_A and Q_B outputs on Plot 1 in the report.

Notice that when an LED is ON, the Q output is HIGH. The red and green LEDs indicate that the pulse generator frequency has been changed by the flip-flops.

For Further Investigation

Measurement of t_{PLH} and t_{PHL}

Note: The measurement of a parameter such as t_{PLH} is done differently for analog and digital scopes. Set up the experiment as in step 1; then if you are using an analog scope, do step 2a. If you are using a digital scope, do step 2b. If both scopes are available, do both steps 2a and 2b.

1. Set up the J-K flip-flop for toggle operation. Set the clock frequency for 100 kHz and view the clock on channel 1 and the Q output on channel 2 of your oscilloscope. Set the scope sweep time for 5 μs/div to observe the complete waveforms of both the clock and the Q output. Set the VOLTS/DIV control on each channel to 2 V/div and center the two waves across the center graticule of the display.

2a. With an analog scope, you will need to trigger the scope on the earlier signal (the clock). Trigger the scope using CH1 and select falling-edge triggering from the trigger controls. Then

---

*The data sheet in Appendix A shows both types.

**Ripple counters are covered further in Experiment 21.

120

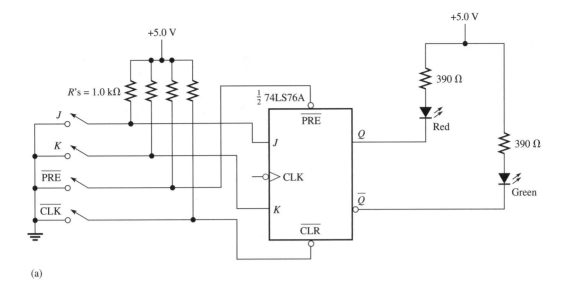

(a)

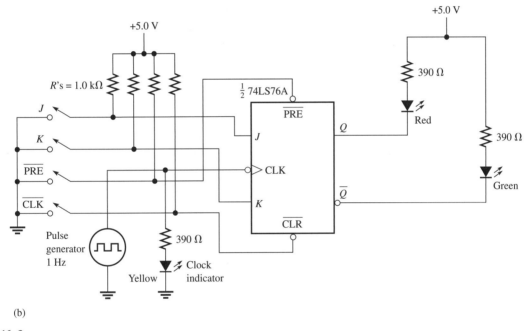

(b)

FIGURE 16–2

increase the sweep speed to 5 ns/div (or use the fastest available sweep time if 5 ns/div is not possible). You may need to adjust the trigger LEVEL control to see the entire clock waveform. You should see a falling edge of the clock and either a rising or falling edge of the Q output. You can observe the LOW-to-HIGH transition of the output by adjusting the HOLDOFF control. When you have a stable trace, go to step 3.

2b. With a digital scope, you can trigger on the slower waveform (the output) which is on channel 2. This is because the DSO can show signals before the trigger event. From the trigger menu, select CH2 triggering and select SET LEVEL TO 50%. Then increase the sweep speed to 5 ns/div. In the trigger menu, you can choose between RISING SLOPE or FALLING SLOPE to observe t_{PLH} or t_{PHL}, respectively.

3. Measure the time from the 50% level of the falling clock signal to the 50% level on the output signal for both a rising and falling output signal. Record your time in the report and compare it to the manufacturer's specified maximum values from the data sheet in Appendix A.

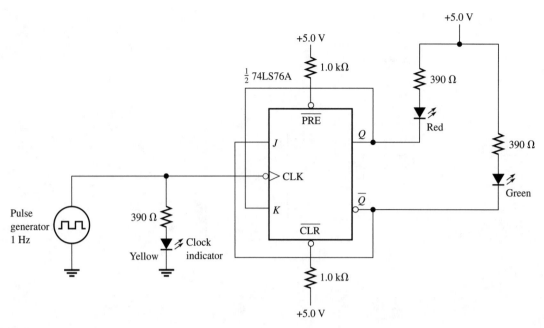

FIGURE 16–3

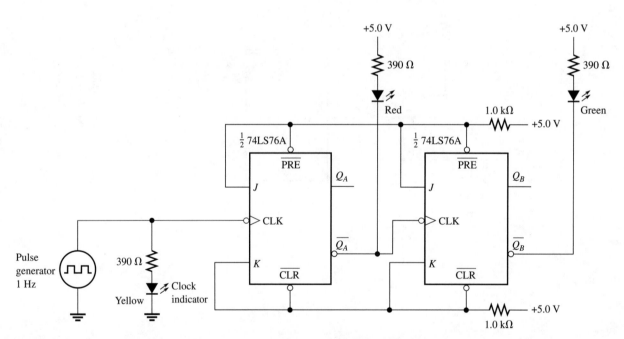

FIGURE 16–4

Report for Experiment 16

Name: _____ Date: _____ Class: _____

Objectives:

- ☐ Test various configurations for a J-K flip-flop.
- ☐ Observe frequency division characteristics in the toggle mode.
- ☐ Measure the propagation delay of a J-K flip-flop.

Data and Observations:

Step 1. Observations for $\overline{PRE}$ and $\overline{CLR}$ inputs:

Step 2. Observations of clocking the J-K flip-flop:

Step 3. Observations of test circuit:

Step 4. Ripple counter:

Clock: ⊓⊓⊓⊓⊓⊓⊓⊓⊓⊓⊓⊓⊓⊓

Q_A

Q_B

PLOT 1

Results and Conclusion:

Further Investigation Results:

Evaluation and Review Questions

1. What is the difference between an asynchronous and a synchronous input?

2. **a.** Describe how you would set a J-K flip-flop asynchronously.

   b. How would you set it synchronously?

3. If both J and K inputs are LOW and $\overline{\text{PRE}}$ and $\overline{\text{CLR}}$ are HIGH, what effect does the clock have on the output of a J-K flip-flop?

4. Assume a student accidentally reversed the J and K inputs on the circuit in Figure 16–3. What effect would be observed?

5. Assume the red LED in Figure 16–3 is on steady and the green LED is off. The yellow LED is blinking. What are three possible troubles with the circuit?

6. Assume the green LED in Figure 16–4 is off but the red LED is blinking. A check at the CLK input of the second flip-flop indicates clock pulses are present. What are possible troubles with the circuit?

CHAPTER 6: LATCHES, FLIP-FLOPS, AND TIMERS

Experiment 17
One-Shots and Astable Multivibrators

Objectives

After completing this experiment, you will be able to

□ Specify components and trigger logic for a 74121 one-shot to produce a specified pulse and trigger mode.

□ Measure the frequency and duty cycle of a 555 timer configured as an astable multivibrator.

□ Specify components for a 555 timer configured as an astable multivibrator and test your design.

Materials Needed

74121 one-shot
7474 dual flip-flop
555 timer
Two 0.01 μF capacitors
Signal diode (1N914 or equivalent)
Resistors: 10 kΩ, 7.5 kΩ
Other components determined by student

Summary of Theory

There are three types of multivibrators: the bistable, the monostable (or *one-shot*), and the astable. The name of each type refers to the number of stable states. The bistable is simply a latch or flip-flop that can be either set or reset and will remain in either state indefinitely. The one-shot has one stable (or inactive) state and one active state, which requires an input trigger to assert. When triggered, the one-shot enters the active state for a precise length of time and returns to the stable state to await another trigger. Finally, the astable multivibrator has no stable state and alternates (or "flip-flops") between HIGH and LOW by itself. It frequently functions as a clock generator, since its output is a constant stream of pulses. Many systems require one-shot or astable multivibrators. The traffic signal control system,[*] requires two one-shots and an astable multivibrator as a clock. In this experiment, you will specify the components for the astable multivibrator and test the frequency and duty cycle. In the For Further Investigation section, you will design the one-shots.

Most applications for one-shots can be met with either an IC timer or an IC one-shot. A timer is a general-purpose IC that can operate as an astable or as a one-shot. As a one-shot, the timer is limited to pulse widths of not less than about 10 μs or frequencies not greater than 100 kHz. For more stringent applications, the IC one-shot takes over. The 74121, which you will test in this experiment, can provide pulses as short as 30 ns. In addition, integrated circuit one-shots often have special features, such as both leading and trailing edge-triggering and multiple inputs that can allow triggering only for specific logic combinations. These can be extremely useful features. The logic circuit and function table for the 74121 are shown in Figure 17–1. The circuit

---

*Discussed in Section 6-1 of the text.

is triggered by a rising pulse on the output of the Schmitt AND gate. The purpose of the Schmitt AND gate is to allow slow rise-time signals to trigger the one-shot. In order for B to trigger it, the input must be a rising pulse, and either A_1 or A_2 must be held LOW, as shown in the last two lines of the function table. If B is held HIGH, then a trailing edge trigger on either A_1 or A_2 will trigger the one-shot provided the other A input is HIGH. Other combinations can be used to inhibit triggering.

This experiment includes an introduction to the 555 timer, the first and still the most popular timer. It is not a TTL device but can operate on +5.0 V (and up to +18 V), so it can be TTL- or CMOS-compatible. This timer is extremely versatile but has limited triggering logic. Some applications include accurate time-delay generation, pulse generation, missing pulse detectors, and voltage-controlled oscillators (VCOs).

Procedure

Monostable Multivibrator Using the 74121

1. The 74121 contains an internal timing resistor of 2.0 kΩ. You can select the internal resistor for the timing resistor by connecting R_{INT} to V_{CC}, or you can select an external resistor. To use an external timing resistor, connect it as shown in Figure 17–1 with R_{INT} (pin 9) left open. The capacitor is an external component but can be eliminated for very short pulses.*

The equation that gives the approximate pulse width t_W is

$$t_W = 0.7 C_{EXT} R_T$$

where R_T is the appropriate timing resistor, either internal or external. C_{EXT} is in pF, R_T is in kΩ, and t_W is in ns. Using a 0.01 µF capacitor, calculate the required timing resistor to obtain a 50 µs pulse width. Obtain a resistor near the calculated value. Measure its resistance and measure the capacitance C_{EXT}. Record the computed R_T and the measured values of R_T and C_{EXT} in Table 17–1 of the report.

2. Using the measured values of R_T and C_{EXT}, compute the expected pulse width, t_W. Record the computed value in Table 17–1.

3. Assume that you need to trigger the one-shot using a leading-edge trigger from the pulse generator. Determine the required connections for $A_1, A_2,$ and B. List the input logic levels and the generator connection in your report. Build the circuit. One-shots are susceptible to noise pickup, so you should install a 0.01 µF bypass capacitor from V_{CC} to ground as close as possible to the 74121.

4. Apply a 10 kHz TTL-compatible signal from the pulse generator to the selected trigger input. Look at the pulse from the generator on channel 1 of a two-channel oscilloscope and the Q output on channel 2. Measure the pulse width and

---

*The 74121 is discussed in detail in Section 6-5 of the text.

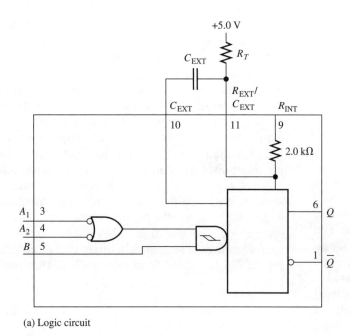

(a) Logic circuit

Inputs			Outputs	
A_1	A_2	B	Q	$\bar{Q}$
L	X	H	L	H
X	L	H	L	H
X	X	L	L	H
H	H	X	L	H
H	↓	H	⎍	⎍
↓	H	H	⎍	⎍
↓	↓	H	⎍	⎍
L	X	↑	⎍	⎍
X	L	↑	⎍	⎍

H = high logic level
L = low logic level
X = can be either low or high
↑ = positive going transition
↓ = negative going transition
⎍ = a positive pulse
⎍ = a negative pulse

(b) Function table

FIGURE 17–1

compare it with the expected pulse width from Step 1. (You may need to adjust R.) Record the measured pulse width in Table 17–1.

5. Increase the frequency slowly to 50 kHz while viewing the output on the scope. What evidence do you see that the 74121 is not retriggerable? Describe your observations.

The 555 Timer as an Astable Multivibrator

6. One of the requirements for many circuits is a clock, a series of pulses used to synchronize the various circuit elements of a digital system. In the astable mode, a 555 timer can serve as a clock generator.

A basic astable circuit is shown in Figure 17–2. There are two timing resistors. The capacitor is charged through both but is discharged only through R_2. The duty cycle, which is the ratio of the output HIGH time t_H divided by the total time T, and the frequency f are found by the following equations:

$$\text{Duty cycle} = \frac{t_H}{T} = \frac{R_1 + R_2}{R_1 + 2R_2}$$

$$f = \frac{1.44}{(R_1 + 2R_2)C_1}$$

Measure the value of two resistors R_1 and R_2 and capacitor C_1 with listed values as shown in Table 17–2. Record the measured values of the components in Table 17–2. Using the equations, compute the expected frequency and duty cycle for the 555 astable multivibrator circuit shown in Figure 17–2. Enter the computed frequency and duty cycle in Table 17–2.

7. Construct the astable multivibrator circuit shown in Figure 17–2. Using an oscilloscope, measure the frequency and duty cycle of the circuit and record it in Table 17–2.

8. With the oscilloscope, observe the waveforms across capacitor C_1 and the output waveform at the same time. On Plot 1, sketch the observed waveforms.

9. While observing the waveforms from Step 8, try placing a short across R_2. Remove the short and write your observations in space provided in the report.

10. A clock oscillator signal, generated from an astable multivibrator, is required for the traffic signal control system. The specified oscillator frequency is 10 kHz. The circuit in Figure 17–2 oscil-

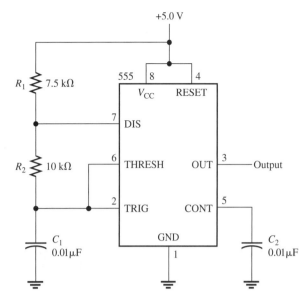

FIGURE 17–2

lates at too low a frequency. Modify the design of this circuit so that it oscillates at 10 kHz (the duty cycle is not critical). Show the circuit in the space provided in the report.

For Further Investigation

The traffic signal control system was shown in block diagram form in Experiment 13 (see Figure 13–4). The system requires two one-shots, shown in the system as the *short timer* and the *long timer*. When the state decoder changes from LOW to HIGH, it causes the trigger logic to change from HIGH to LOW (trailing edge). It is this HIGH-to-LOW transition (trailing edge) that is used to trigger the timers. The short timer must have a 4 s positive pulse and the long timer must have a 25 s positive pulse. Check the manufacturer's maximum values of R_T and C_{EXT} for the 74121.* Then design and build the circuits. An LED (with 330 Ω current-limiting resistor) can be used as an indicator. Test your design to see that the pulse width is approximately correct. Draw the circuits in the report and indicate your test results.

---

*See Appendix A.

Report for Experiment 17

Name: _____ Date: _____ Class: _____

Objectives:

☐ Specify components and trigger logic for a 74121 one-shot to produce a specified pulse and trigger mode.
☐ Measure the frequency and duty cycle of a 555 timer configured as an astable multivibrator.
☐ Specify components for a 555 timer configured as an astable multivibrator and test your design.

Data and Observations:

TABLE 17–1
Data for 74121 monostable multivibrator.

Quantity	Computed Value	Measured Value
Timing Resistor, R_T		
External Capacitor, C_{EXT}	0.01 µF	
Pulse Width, t_W		

Step 3. Input logic levels and generator connection:

Step 5. Observations as frequency is raised to 50 kHz:

TABLE 17–2
Data for 555 timer as an astable multivibrator.

Quantity	Computed Value	Measured Value
Resistor, R_1	7.5 kΩ	
Resistor, R_2	10.0 kΩ	
Capacitor, C_1	0.01 µF	
Frequency		
Duty Cycle		

Step 8:

Capacitor waveform: | | | | | | | | | | |

Output waveform: | | | | | | | | | | |

PLOT 1

Step 9. Observations with a short across R_2:

Step 10. Circuit for a 10 kHz oscillator for traffic signal controller:

Results and Conclusion:

Further Investigation Results:

Evaluation and Review Questions

1. What does the term *nonretriggerable* mean for a monostable multivibrator?

2. a. For the 74121 monostable multivibrator circuit, compute the value of the capacitor for a pulse width of 50 μs using the internal resistor.

   b. How would you design a monostable multivibrator circuit with a variable output that can be adjusted from 50 μs to 250 μs?

3. From the data sheet for the 74121, determine the largest timing resistor and capacitor recommended by the manufacturer. What pulse width would you predict if these values were chosen for the circuit in Figure 17–1?

$$t_W = \underline{\hspace{2cm}}$$

4. Compute the duty cycle and frequency for a 555 astable multivibrator if $R_1 = 1.0 \text{ k}\Omega$, $R_2 = 180 \text{ k}\Omega$, and $C_1 = 0.01 \text{ μF}$.

5. For the 555 astable multivibrator, determine R_1 and R_2 necessary for a period of 12 s if $C_1 = 10 \text{ μF}$ and the required duty cycle is 0.60.

$$R_1 = \underline{\hspace{2cm}}$$

$$R_2 = \underline{\hspace{2cm}}$$

6. Assume the 555 astable multivibrator in Figure 17–2 is operating from +15 V. What voltage range do you expect to see across capacitor C_1?

Experiment **18**
Shift Register Counters

Objectives

After completing this experiment, you will be able to
- Test two recirculating shift register counters.
- From oscilloscope measurements, draw the timing diagram for the two shift register counters.

Materials Needed

74195 4-bit shift register
7400 quad NAND gate
7493A counter
7474 D flip-flop
7486 quad exclusive OR
Four-position DIP switch
Four LEDs
Resistors: four 330 Ω, six 1.0 kΩ
Two N.O. pushbuttons (optional)

Summary of Theory

A *shift register* is a series of flip-flops connected so that data can be transferred to a neighbor each time the clock pulse is active. An example is the display on your calculator. As numbers are entered on the keypad, the previously entered numbers are shifted to the left. Shift registers can be made to shift data to the left, to the right, or in either direction (bidirectional), using a control signal. They can be made from either D or J-K flip-flops. An example of a simple shift register made from D flip-flops is shown in Figure 18–1(a). The data are entered serially at the left and may be removed in either parallel or serial fashion. With some additional logic, the data may also be entered in parallel, as shown in Figure 18–1(b).

Shift registers are available in IC form with various bit lengths, loading methods, and shift directions. They are widely used to change data from serial form to parallel form, and vice versa. Other applications for shift registers include arithmetic operations in computers. To multiply any number by its base, you simply move the radix point one position to the left. To multiply a binary number by 2, the number is shifted to the left. For example, $7 \times 2 = 14$ in binary is $0111 \times 10 = 1110$. Note that the original number 0111 is shifted by one position to the left. Conversely, division by 2 is represented by a right shift.

Another application of the shift register is as a digital waveform generator. Generally, a waveform generator requires feedback—that is, the output of the register is returned to the input and recirculated. Two important waveform generators are the Johnson (or "twisted-ring") counter and the ring counter. The names can be easily associated with the correct circuit if the circuits are drawn in the manner shown in Figure 18–2. In this experiment, you will construct both of these counters using a 74195 4-bit shift register. The ring counter will then be used to generate a bit stream that can be used in different systems.

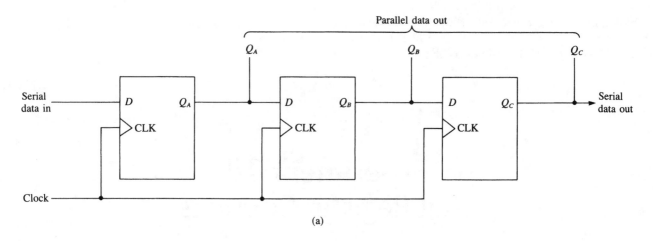

(a)

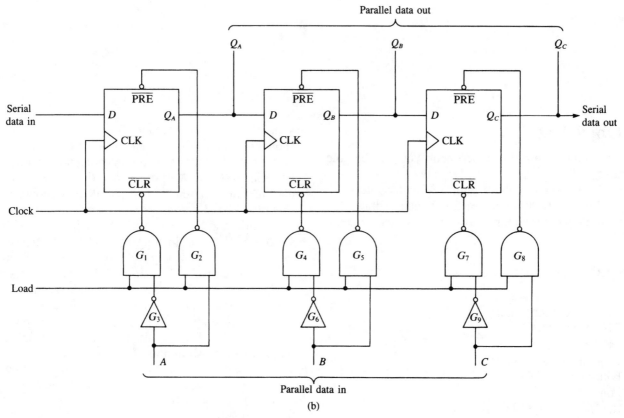

(b)

FIGURE 18–1
Shift registers made from D flip-flops.

The 74195 function table is in the manufacturer's data sheet* and is reproduced in Table 18–1 for convenience. The first input listed on the table is an asynchronous $\overline{\text{CLEAR}}$. Next is a parallel SHIFT/$\overline{\text{LOAD}}$ function on one pin. Assertion level logic is shown to define that a HIGH causes the register to SHIFT from Q_A toward Q_D at the next clock edge, and a LOW causes the register to $\overline{\text{LOAD}}$ at the next clock edge. The inputs A through D are used only when the register is loaded in parallel (also called a *broadside load*). Notice that the internal register portion of the 74195 is shown with S-R flip-flops, but the serial inputs to the leftmost flip-flop are labeled as J and $\overline{K}$. These inputs function the same as the inputs to an ordinary J-K flip-flop, except the K input is inverted.

*See Appendix A.

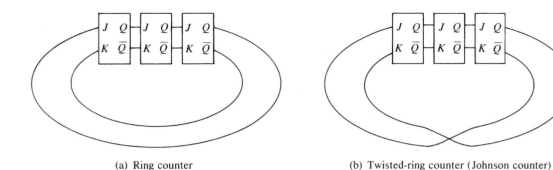

(a) Ring counter   (b) Twisted-ring counter (Johnson counter)

FIGURE 18–2
Shift register counters drawn to emphasize their names. These circuits were drawn with J-K flip-flops but can be constructed from other flip-flops as well. The CLK, $\overline{PRE}$, and $\overline{CLR}$ inputs are not shown.

TABLE 18–1
Function table for 74195 4-bit shift register.

Inputs									Outputs				
	SHIFT/		SERIAL		PARALLEL								
$\overline{CLEAR}$	$\overline{LOAD}$	CLOCK	J	$\overline{K}$	A	B	C	D	Q_A	Q_B	Q_C	Q_D	$\overline{Q}_D$
L	X	X	X	X	X	X	X	X	L	L	L	L	H
H	L	↑	X	X	a	b	c	d	a	b	c	d	$\overline{d}$
H	H	L	X	X	X	X	X	X	Q_{A0}	Q_{B0}	Q_{C0}	Q_{D0}	$\overline{Q}_{D0}$
H	H	↑	L	H	X	X	X	X	Q_{A0}	Q_{A0}	Q_{Bn}	Q_{Cn}	$\overline{Q}_{Cn}$
H	H	↑	L	L	X	X	X	X	L	Q_{An}	Q_{Bn}	Q_{Cn}	$\overline{Q}_{Cn}$
H	H	↑	H	H	X	X	X	X	H	Q_{An}	Q_{Bn}	Q_{Cn}	$\overline{Q}_{Cn}$
H	H	↑	H	L	X	X	X	X	$\overline{Q}_{An}$	Q_{An}	Q_{Bn}	Q_{Cn}	$\overline{Q}_{Cn}$

H = high level (steady state)

L = low level (steady state)

X = irrelevant (any input, including transitions)

↑ = transition from low to high level

a, b, c, d = the level of steady state input at A, B, C, or D, respectively

Q_{A0}, Q_{B0}, Q_{C0}, Q_{D0} = the level of Q_A, Q_B, Q_C, or Q_D, respectively, before the indicated steady-state input conditions were established

Q_{An}, Q_{Bn}, Q_{Cn} = the level of Q_A, Q_B, or Q_C, respectively, before the most recent transition of the clock

Procedure

Johnson and Ring Counters

1. The circuit shown in Figure 18–3 is a partially completed schematic for a shift register counter. It could be connected as either a Johnson (twisted-ring) or as a ring counter. Refer to Figure 18–2 and the function table for the 74195 (Table 18–1) and determine how to complete the feedback loop for a twisted-ring counter. Show the completed schematic in your report.

2. Connect your circuit. (The $\overline{CLEAR}$ and SHIFT/$\overline{LOAD}$ pushbuttons can be made with pieces

of hook-up wire.) Set the pulse generator for a TTL pulse at 1 Hz. Momentarily close the $\overline{CLEAR}$ switch. One useful feature of the counter is that when the sequence begins in state 0, it forms a Gray code sequence. Although you could load a pattern other than all zeros, this is a typical starting point for a Johnson counter.

3. Observe the pattern in the LEDs. (The LEDs are ON for a zero.) Then speed up the pulse generator to 1 kHz and develop a timing diagram for the Johnson counter outputs. Draw your timing diagram in the space provided in the report.

4. Referring to Figure 18–2 and the function table for the 74195, change the schematic to that of

FIGURE 18–3
Partially completed schematic for
twisted-ring or ring counter.

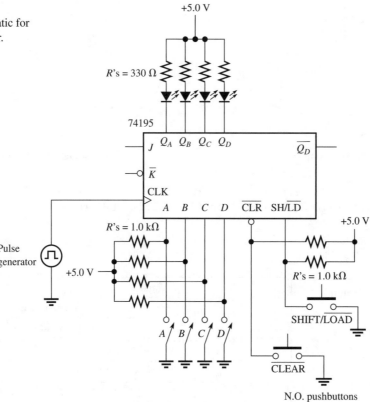

a ring counter. The partially completed schematic is shown in the report. A ring counter does not invert the bits that are fed back, so the desired bit pattern must be preset through the parallel load feature of the shift register. A common pattern is to have either a single 1 or a single 0 recirculate. Set the load switches for 1110_2 and press the SHIFT/LOAD pushbutton. From the function table, note that this is a synchronous load, so loading will take place only if a clock is present.

5. Reset the pulse generator for 1 Hz and observe the pattern in the LEDs.* After observing the pattern in the LEDs, speed up the pulse generator to 1 kHz and develop a timing diagram for the ring counter outputs. Draw your timing diagram in the space provided in the report.

For Further Investigation

This investigation is a bit different than previous ones. In this investigation, it is not necessary to build the circuit; rather you should try to figure out

timing details (of course you could build it if you choose). The circuit is an automated IC tester for 2-input gates shown in Figure 18–4. It uses a 74195 shift register to generate a serial data train that represents the predicted data for a device under test (D.U.T.). The way it works is that a 2-input gate receives four states from the 7493A counter and produces a logical one or zero depending on the type of gate that is tested. If the data from the D.U.T. matches the shift register data, the test continues; otherwise the *Device failed LED* will come on. Timing for this simple system is not trivial but by carefully drawing the waveforms for each stage, you can figure out how it works. Start by drawing the waveforms for the 7493A. Assume a 2-input NAND gate is the D.U.T. and the predict data is set for $A = 0$, and $B = C = D = 1$. Show the time relationship between the counter and the Strobe, Input test data, and the Serial predict data. Summarize in a short report how the circuit works and what happens if the Input test data doesn't match the Serial predict data.

---

*This pattern is essentially the same pattern used in the ring counter for the keyboard encoder shown in Figure 7–28 of Floyd's text.

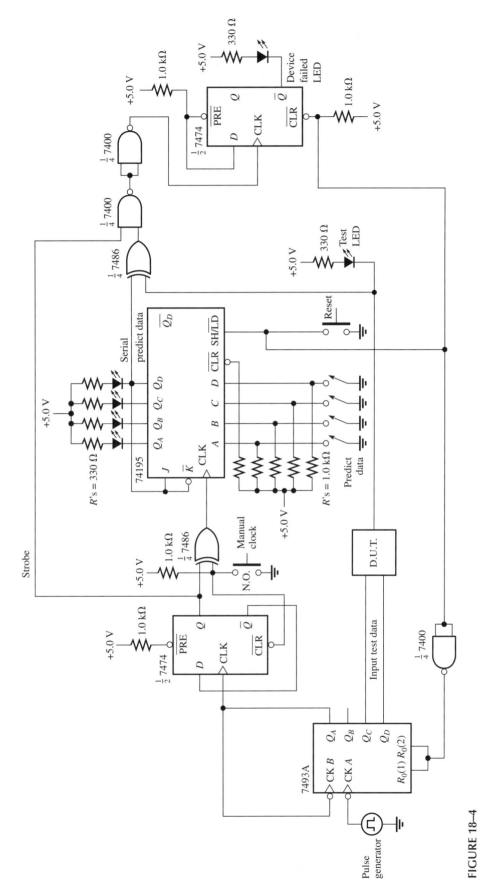

FIGURE 18–4
Automated IC tester.

Report for Experiment 18

Name: _____ Date: _____ Class: _____

Objectives:

☐ Test two recirculating shift register counters.
☐ From oscilloscope measurements, draw the timing diagram for the two shift register counters.

Data and Observations:

Schematic for Johnson counter: Schematic for ring counter:

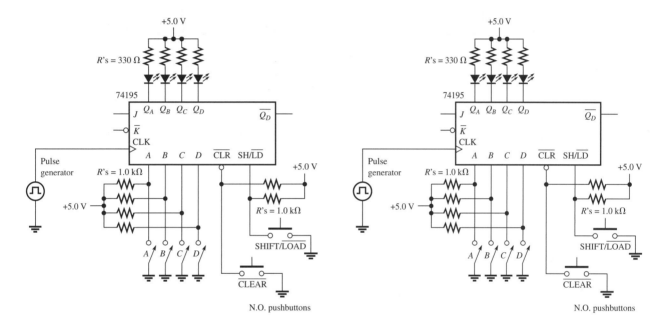

Timing diagram for
Johnson counter:

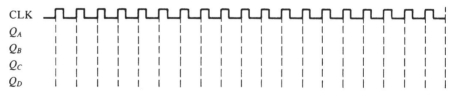

Timing diagram for ring
counter loaded with 1110:

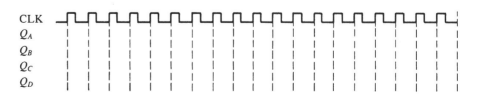

Results and Conclusion:

Further Investigation Results:

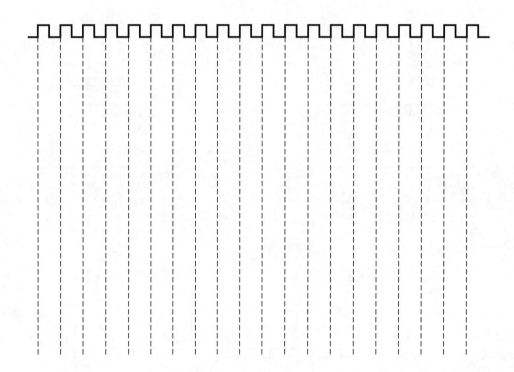

Evaluation and Review Questions

1. How would you connect a second 74195 to extend the Johnson counter to 8 bits?

2. Explain why it is necessary to use an edge-triggered device for a shift register.

3. a. A 3-stage ring counter is loaded with the binary number 101. What are the next three states of the counter?

   b. Repeat (a) for a Johnson counter.

4. Ring counters and Johnson counters both have a feedback path from the output to the input. If the feedback path opens for either counter, what happens to all outputs?

5. Assume the outputs of the Johnson counter that you drew in the report were decoded to show a series of unique codes. Draw the expected waveforms you would observe on a logic analyzer.

6. Assume your boss asks you to build a decoder for a ring counter. What do you say to him? (be polite!).

Experiment 19
Application of Shift Register Circuits

Objectives

After completing this experiment, you will be able to

□ Construct an asynchronous data transmitter and use an oscilloscope to observe and plot the asynchronous output signal.

□ Design a circuit to generate exactly 5 clock pulses for the data receiver.

□ Complete the design of an asynchronous data receiver.

Materials Needed

555 timer
7474 dual D flip-flop
74195 shift register
Six LEDs
Capacitors: one 0.1 µF, one 1.0 µF
Resistors: six 330 Ω, eight 1.0 kΩ, one 10 kΩ, one 22 kΩ, one 1.0 MΩ
Other materials as determined by student

For Further Investigation:
7400 quad 2-input NAND gate
7493A counter
Additional 74195 shift register
Four LEDs
Resistors: four 330 Ω, three 1.0 kΩ

Summary of Theory

In most digital systems, data is moved and processed in parallel form because this form is fast.

When data must be transferred to another system, it is frequently sent in serial form because serial form requires only a single data line. A requirement of such systems is the conversion of data from parallel form to serial form and back. A widely used method of sending serial data is called *asynchronous* data transmission. In asynchronous transmission, the bits are sent at a constant rate, but there are varying spaces between groups of bits. To identify a new group, a start bit is sent, which is always LOW. This is followed by the data bits (usually 8) and one or two stop bits, which are always HIGH. The receiver is idle until a new start bit is received. This sequence is illustrated in Figure 19–1.

The preceding system will be simplified in this experiment to illustrate the process of sending and receiving asynchronous data. The transmitter shown in Figure 19–2 is simply a 4-bit 74195 shift register that has been extended to 6 bits by the addition of two D flip-flops. The 74195 is first loaded with the four data bits to be sent. The D flip-flops are used for the start bit and to keep the serial data line HIGH as indicated. The one-shot debounces the SHIFT/$\overline{\text{LOAD}}$ line.

Part of this experiment requires you to design an asynchronous receiver. In the For Further Investigation section, you can construct and test the receiver you designed. Although the transmitter and receiver in this experiment illustrate the concepts of data conversion and transmission, they will not operate properly over a distance of more than a few feet. Transmission over greater distances requires

143

FIGURE 19–1
Asynchronous data transmission.

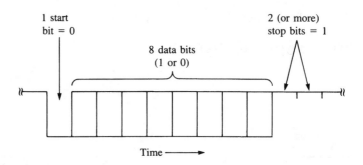

the use of special circuits called *line drivers* and *line receivers,* usually using special cable or twisted pair wiring.

Procedure

1. The asynchronous data transmitter circuit shown in Figure 19–2 sends one start bit and four data bits. This requires a 6-bit register because the serial data line must be held HIGH during periods

when no data is sent. The receiver (not shown) waits for the HIGH to LOW transition of the start bit and then clocks the data bits into a shift register. Construct the transmitter shown in Figure 19–2. The 555 timer is configured as a one-shot with approximately a 1 s pulse that is triggered by the $\overline{\text{LOAD}}$ switch closure.

2. Test the circuit by setting the data switches for an arbitrary pattern. Set the pulse generator for 1 Hz and press the $\overline{\text{LOAD}}$ pushbutton. As soon as the data is loaded, it should immediately begin to

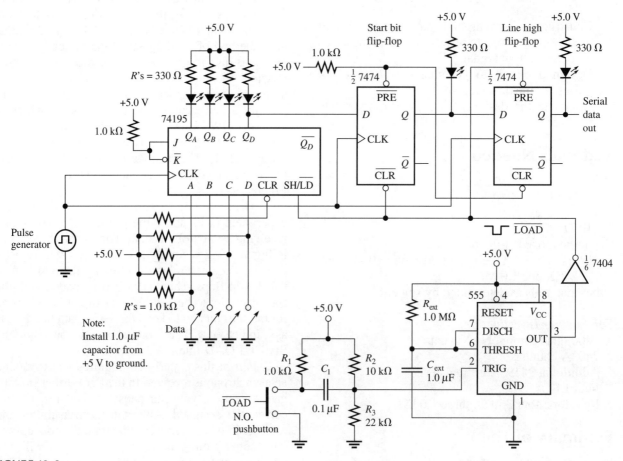

FIGURE 19–2
Asynchronous data transmitter.

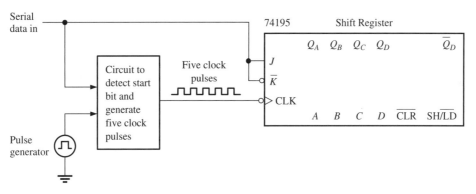

FIGURE 19–4
Block diagram of asynchronous data receiver.

shift through the register. Try some different data patterns and observe the results. Note that the LEDs are ON when the data is LOW. Summarize your observations in the space provided in the report.

    3. In actual data transmission systems, the data is sent considerably faster. You can speed up the process if you automatically reload the data after each transmission and set the pulse generator for a higher frequency. This will also make it easier to observe the data on an oscilloscope. Set the pulse generator to a frequency of 10 kHz. To automatically load the data, change the 555 to an astable multivibrator with a frequency of 1.0 kHz and a duty cycle of approximately 80%.* Show the changes on the partially completed schematic (Figure 19–3) in the report. In particular, show what to do with the Discharge, Threshold, and Trigger inputs of the 555. Note that the inverter is no longer needed.

    4. Observe the pattern of the asynchronous data from the transmitter using an oscilloscope or logic analyzer. Set the data to $DCBA = 0101$. You will need a stable display on the oscilloscope to observe this data as it is repeatedly sent. When you have obtained a stable pattern on the scope, draw the pattern in the report and show the timing relationship between the serial data out and the SHIFT/$\overline{\text{LOAD}}$ signals.

    5. Complete the design of the receiver. The receiver side of the system must generate exactly five clock pulses when it detects the HIGH to LOW transition of the start bit. The five clock pulses are connected to the clock input of the 74195 to shift the serial data into the shift register as shown in the

block diagram of the receiver in Figure 19–4. The block diagram is expanded into a partially completed schematic in Figure 19–5, shown in the report. Complete the design for this circuit. In particular, show what to do with the J and $\overline{K}$, $\overline{\text{CLR}}$, and SH/$\overline{\text{LD}}$ inputs of the 74195 shift register, the inputs to both NAND gates, and CK A of the 7493A counter.

For Further Investigation

Construct the receiver you designed in Step 5. Connect the serial data from the transmitter to the receiver and send the data. Remember, the Q_A output of the 7493A counter divides the pulse generator frequency by two. To ensure that the receiver is clocking the data at the same rate as the transmitter, the receiver pulse generator frequency will need to be set to *twice* the transmitter pulse generator frequency. Set the transmitter pulse generator for 10 kHz and the receiver pulse generator for 20 kHz. Summarize your observations in the space provided in the report.

Multisim Troubleshooting (Optional)

The companion website for this manual has Multisim 11 and 12 files. Download the file named Exp-19nf and the worksheet Exp-19ws. Open the file named Exp-19nf. To make the simulation work at a reasonable speed, the external resistor labeled R_{ext} is smaller in the simulation than in the lab circuit and the pushbutton is replaced with a SPST switch. The shift register is loaded when J1 is closed and opened; the X1 probe will indicate the load operation and shifting occurs as soon as this pulse goes HIGH. Open Exp-19nf to analyze the circuit. Complete the worksheet and attach it to the report.

---

*See Section 6–5 of Floyd's text.

Report for Experiment 19

Name: _____ Date: _____ Class: _____

Objectives:

□ Construct an asynchronous data transmitter and use an oscilloscope to observe and plot the asynchronous output signal.
□ Design a circuit to generate exactly 5 clock pulses for the data receiver.
□ Complete the design of an asynchronous data receiver.

Data and Observations:

Observations from Step 2:

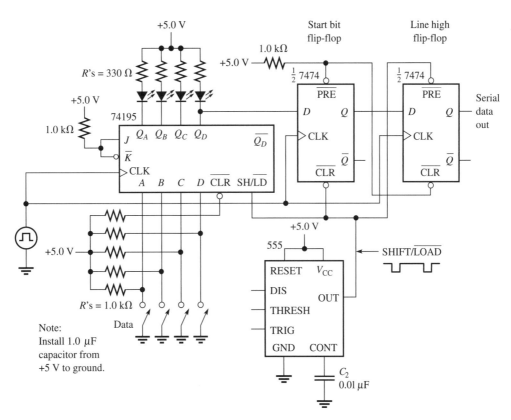

FIGURE 19–3

Step 4: Timing diagram from serial data out line and SHIFT/$\overline{\text{LOAD}}$ signal.

Serial data out

SHIFT/$\overline{\text{LOAD}}$

Step 5: Receiver schematic (partially completed).

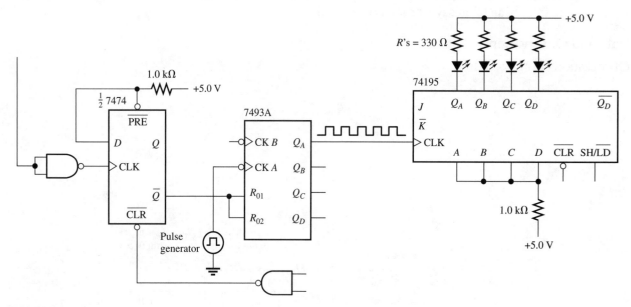

FIGURE 19–5

Results and Conclusion:

Further Investigation Results:

Evaluation and Review Questions

1. After the transmitter (Figure 19–2) has sent the data, the shift register is loaded with all ones. Explain why this occurs.

2. There were only four data bits, yet the receiver was designed to clock *five* bits into the shift register. Explain.

3. Why is the receiver in Figure 19–5 shown with a NAND gate operating as an inverter on the clock input of the 7474?

4. **a.** Figure 19–6 shows how shift registers can be connected with a single full-adder to form a serial adder. Analyze the circuit and explain how it operates.

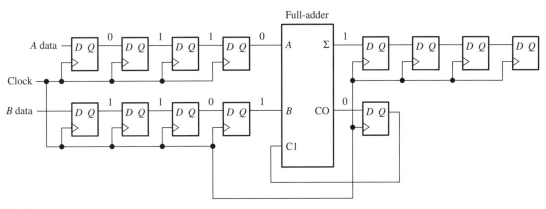

FIGURE 19–6

   b. Assume the input shift registers contain the data as shown. What are the contents of the output shift register and the carry-out flip-flop after four clock pulses?

5. Using a 4-bit shift register, design a circuit that produces the two-phase clock signals shown in Figure 19–7.

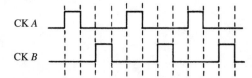

FIGURE 19–7

6. What failures could account for a transmitter that does not send a start pulse but does send the data?

Experiment 20
The Baseball Scoreboard

Objectives

After completing this experiment, you will be able to

☐ Design and build a logic circuit for part of a baseball scoreboard using shift registers or counters.

☐ Write a formal laboratory report describing the circuit and results.

Materials Needed

Five LEDs (two red, three green)
Two N.O. pushbuttons
One 555 timer
Resistors: one 1.0 kΩ, one 10 kΩ, one 22 kΩ, one 1.0 MΩ
Capacitors: one 0.01 µF, one 0.1 µF
Other materials as determined by student

Summary of Theory

A counter uses flip-flops to "remember" the last count until the clock pulse occurs. A shift register can store several events in series depending on the length of the register. A shift register, such as a Johnson counter, can be set with a characteristic sequence like a counter and can be used in applications requiring a counter.

In this experiment, we require two count sequences. Either sequence may be completed first. The sequence completed first must clear both counting devices. One approach might be to use two sep-

arate decoders, either of which can clear the counters. The method of solution is entirely up to you. The counting devices are "clocked" by manual pushbuttons. The pushbuttons will need to be debounced. One method of doing this is shown in Figure 20–1.

Procedure

The Baseball Scoreboard

Design a circuit that solves the baseball scoreboard problem stated next. There are two inputs, controlled by pushbutton switches (normally open contacts). There are five outputs represented by LEDs. Build the circuit. Summarize your design steps and results in a formal lab report.

Problem Statement: The Latchville Little League needs a new baseball scoreboard (see Figure 20–2.) Your assignment is to do the logic for the strikes-and-balls display. The scoreboard will have two lights (LEDs) for strikes and three lights for balls. The scoreboard operator will have two pushbuttons: one for strikes and the other for balls. Each press of the strike pushbutton turns on one more of the strike lights unless two are already on. If two are on, all lights, including strikes and balls, are cleared. (Note that the count sequence for the strike lights is a binary 00-01-11-00.) The balls pushbutton works in a similar manner. Each press causes one more light to come on unless three are already on. If three lights are already on, then all lights, including strikes, are cleared.

FIGURE 20–1

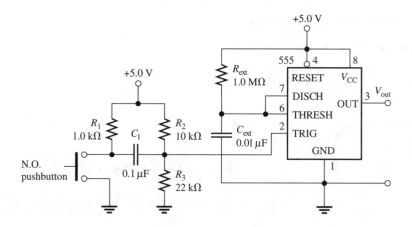

Design Hint: Consider using a 74175 IC for the counting device. The complete circuit can be designed to fit on a single 47-row protoboard.

Report

Write a technical report summarizing your circuit and test results. Your instructor may have a required format, or you may use the one given in the Introduction to the Student.

For Further Investigation

Design the logic circuits needed to complete the scoreboard as illustrated in Figure 20–2. The inning is indicated by a single light (show this as an LED), which corresponds to the inning number. The inning display is controlled by a single pushbutton to advance the light. The outs display is indicated by two lights, which are controlled by a single pushbutton. When the third out is pressed, all lights on the lower row of the scoreboard are cleared.

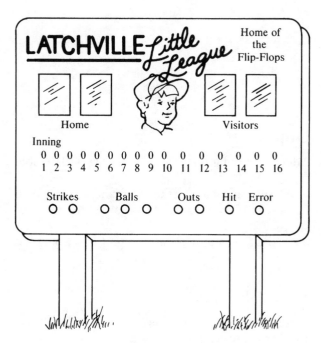

FIGURE 20–2

Experiment 21
Asynchronous Counters

Objectives

After completing this experiment, you will be able to
□ Build and analyze asynchronous up and down counters.
□ Change the modulus of a counter.
□ Use an IC counter and truncate its count sequence.

Materials Needed

7400 quad NAND gates
7474 dual D flip-flop
7493A binary counter
Two LEDs
Resistors: two 330 Ω, two 1.0 kΩ

For Further Investigation:
   7486 quad XOR gate

Summary of Theory

Digital counters are classified as either *synchronous* or *asynchronous,* depending on how they are clocked. Synchronous counters are made from a series of flip-flops that are clocked together. By contrast, asynchronous counters are a series of flip-flops, each clocked by the previous stage, one after the other. Since all stages of the counter are not clocked together, a "ripple" effect propagates as various flip-flops are clocked. For this reason, asynchronous counters are called *ripple counters*. You can easily make a ripple counter from D or J-K flip-flops by connecting them in a toggle mode.

The *modulus* of a counter is the number of different output states the counter may take. The counters you will test in the first four steps of this experiment can represent the numbers 0, 1, 2, and 3; therefore, they have a modulus of 4. You can change the modulus of a ripple counter by decoding any output state and using the decoded state to asynchronously preset or clear the current count. Ripple counters can be made to count either up or down. (They can be made to count both up and down, but usually it is easier to use a synchronous counter for an up/down counter.)

Two methods for changing a counter from up to down or vice versa are illustrated in this experiment. The first method involves moving the logical "true" output of the counter to the other side (as illustrated in Figures 21–2 and 21–3). The second method changes the manner in which the counter is triggered.

If we tabulate a binary count sequence, we note that the LSB (least significant bit) changes at the fastest rate and the rate of change is divided by 2 as we look at succeeding columns. A typical 3-stage counter might have output waveforms as shown in Figure 21–1. For this counter, we can assign each output with a "weight" equal to the column value that would be assigned to binary numbers. Output

FIGURE 21–1

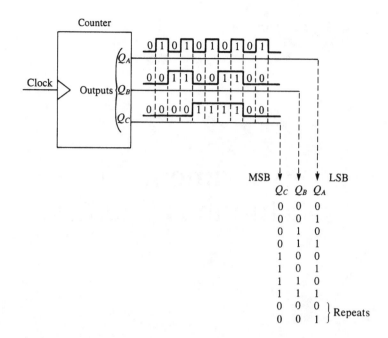

Q_A has a weight of 1, output Q_B has a weight of 2, and output Q_C has a weight of 4. For the counter shown, the count sequence is for an up counter.

Because each stage of a ripple counter changes state at a slightly different time, the counter has a tendency to produce "glitches"— spikes of short duration—when the outputs are decoded owing to the short time delays when the flip-flops are changing states. Glitches are a disadvantage of ripple counters for many applications. Another disadvantage is the limited speed owing to the cumulative delays through the counter. For some applications, such as digital clocks, the slower speed is not a problem.

For most applications requiring a counter, MSI counters are available. The 7493A is an example of an asynchronous counter containing four J-K flip-flops, with the J and K inputs internally wired HIGH, putting them in the toggle mode. Three of the flip-flops are connected together as a 3-bit counter. The fourth flip-flop is separate, including its own clock input. To form a 4-bit counter, connect the Q_A output of a single J-K flip-flop to the clock B input of the 3-bit counter. A common reset line goes to all flip-flops. This reset line is controlled by an internal 2-input NAND gate. You can select any count sequence up to 16 by choosing the internal counter, detecting the desired count, and using it to reset the counter. In the For Further Investigation section, you will be introduced to an idea for changing the up/down count sequence using a control signal.

In this experiment and other experiments with counters, it is necessary to determine the time relationships between various digital signals. If your laboratory is equipped with a logic analyzer*, you may want to use it to capture data from the counters. The basic logic analyzer is a versatile digital instrument that allows you to capture multiple channels of digital data, store them, manipulate them, and view them. A basic logic analyzer converts the input data on each channel to a series of 1's and 0's of digital information and stores them in a digital memory. The data can be stored using an internal or external clock to sample the data at specific time intervals, or the signals can be sampled using an asynchronous signal, such as might be found in an asynchronous data transmission system. After the data are sampled, they can be viewed in several modes, depending on the analyzer. The primary viewing mode is either a set of reconstructed digital waveforms or a state listing of the data in memory. This list can be presented in various formats. More complex analyzers include a built-in digital storage oscilloscope (DSO).

Other important features make logic analyzers important instruments for testing and troubleshooting digital circuits. Because logic analyzers differ in features and capabilities, it is not possible in this summary to explain detailed operation. Refer to the operator's manual for your analyzer for operating instructions.

---

*The logic analyzer is introduced in Section 1–8 of the text.

Procedure

Two-Bit Asynchronous Counters

Note: The LED indicators in this experiment are connected through NAND gates wired as inverters to form the display portion of the circuit. Although not strictly necessary, this method enables you to more easily visualize the ideas presented in the experiment without violating the I_O specification for the 7474. Also, in this experiment, it is necessary to show the difference between the *electrical* output, *Q,* and the *logic* output. Accordingly, the logic output is labeled with the letter *A* or *B*. It is possible for the electrical and logic outputs to have opposite meanings, as will be seen.

1. Construct the 2-bit asynchronous counter shown in Figure 21–2. Clock flip-flop *A* using a 1 Hz TTL pulse from the function generator to the clock input, and watch the sequence on the LEDs. Then speed up the generator to 1 kHz, and view the *A* and *B* output waveforms on a dual-channel oscilloscope. Trigger the oscilloscope from channel 1 while viewing the *B* signal on channel 1 and the *A* signal or the clock on channel 2. Triggering on the slower *B* signal will give a stable trace and assures there is no ambiguity in determining the timing diagram (possible on analog scopes). Sketch the output timing diagram in Plot 1 of the report.

Notice that the frequency of the *B* output is one-half that of the *A* output. As explained in the Summary of Theory section, the column weight of flip-flop *B* is twice that of flip-flop *A*, and thus it can be thought of as the MSB of the counter. By observation of your waveforms, determine whether this is an up counter or a down counter, and record your answer.

2. Now we will change the way we take the "true" output from the counter and see what happens. If logic "truth" is taken from the other side of each flip-flop, then we have the circuit shown in Figure 21–3. Modify your circuit and view the output waveform from each stage. Sketch the timing diagram on Plot 2 of the report.

3. Next we will change the manner in which flip-flop *B* is clocked. Change the circuit to that of Figure 21–4. The "true" output of the counter remains on the $\overline{Q}$ side of the flip-flops. View the outputs as before, and sketch the waveforms on Plot 3.

4. Now change the logic "true" side of the counter, but do not change the clock, as illustrated in Figure 21–5. Again, sketch the outputs of each flip-flop on Plot 4.

5. You can change the modulus of the counter by taking advantage of the asynchronous clear ($\overline{CLR}$) and asynchronous preset ($\overline{PRE}$) inputs of the 7474. Look at the circuit of Figure 21–6, a modification of the circuit in Figure 21–5. Predict the behavior of this circuit, and then build the circuit. Sketch the output waveforms of each flip-flop on Plot 5 and determine the count sequence. Set the generator clock

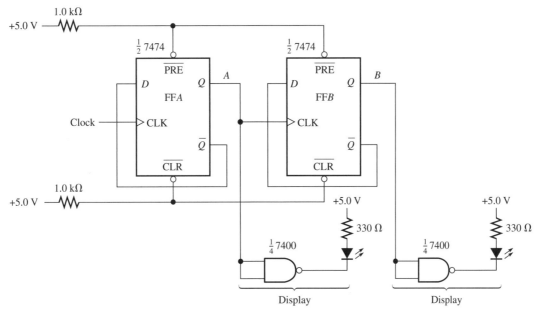

FIGURE 21–2
Ripple counter with D flip-flops.

155

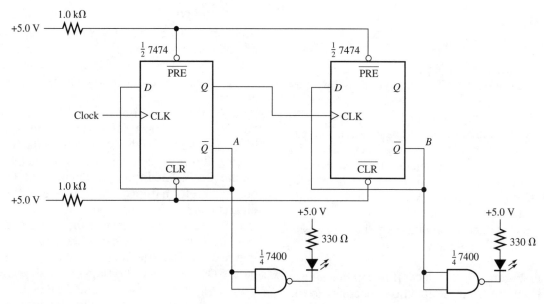

FIGURE 21–3
Ripple counter with D flip-flops. Note that the "true" output is shown on the $\overline{Q}$ output.

frequency at 500 kHz and look for the very short spike that causes the count sequence to be truncated.

6. The very short spike, called a *glitch,* on the *A* output is necessary to cause the counter to reset. While this signal serves a purpose, glitches in digital systems are often troublesome. Let's look at an undesired glitch caused by two flip-flops changing states at nearly the same time.

Add a 2-input NAND gate to the circuit of Figure 21–6, which decodes state $\overline{0}$. (Connect the inputs of the NAND gate to $\overline{A}$ and $\overline{B}$.) Leave the

generator frequency at 500 kHz. Look carefully at the output of the NAND gate. Sketch the observed waveforms on Plot 6.

The 7493A Asynchronous Counter

7. You can configure the 7493A 4-bit binary counter to count from 0 to 15 by connecting the output of the single flip-flop (Q_A) to the clock *B* input. Connect the output of a TTL-level pulse from the function generator to the clock *A* input. From the

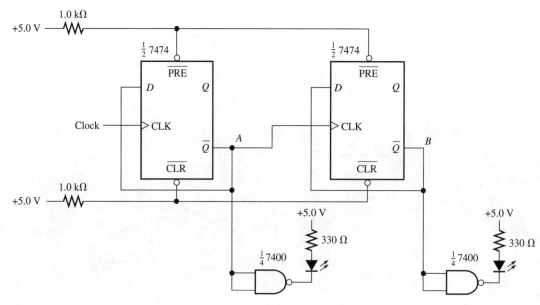

FIGURE 21–4
Ripple counter with D flip-flops. Note that the *B* counter is triggered from the $\overline{Q}$ output.

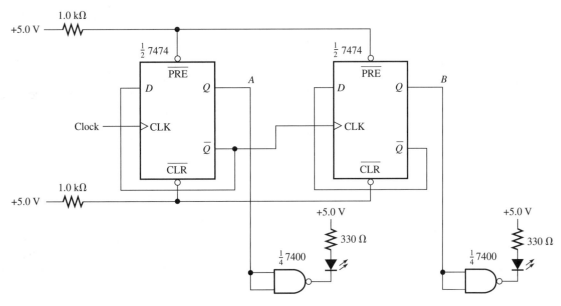

FIGURE 21–5
Ripple counter with D flip-flops. Note that the "true" output is on the Q outputs.

data sheet, determine the necessary connections for the reset inputs.

Set the input frequency for 400 kHz. Trigger a two-channel oscilloscope from the lowest-frequency symmetrical waveform (Q_D), and observe, in turn, each output on the second channel. (If you have a logic analyzer, you can observe all four outputs together.) Sketch the timing diagram on Plot 7.

8. Figure 21–7 shows a 7493A counter configured with a truncated count sequence. Modify your previous circuit and observe the output waveforms on an oscilloscope or logic analyzer. Again, trigger the oscilloscope on the lowest-frequency symmetrical waveform, and observe each output on the second channel. Sketch the timing diagram on Plot 8.

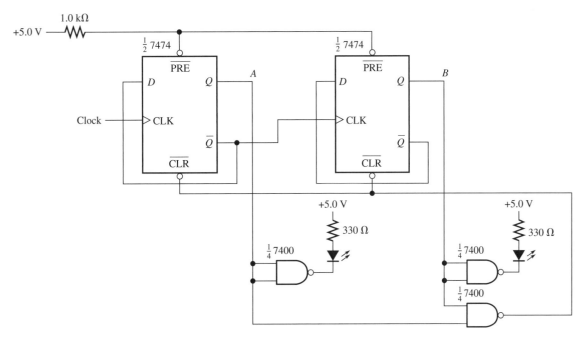

FIGURE 21–6
Ripple counter with D flip-flops and truncated count.

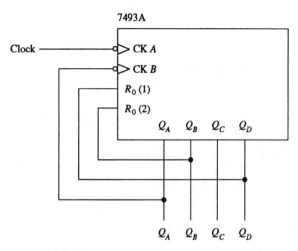

FIGURE 21–7
7493A with truncated count sequence.

For Further Investigation

Adding the UP/DOWN Function to the 7493A

The methods you have investigated so far involve hardware changes to the circuit to change the counter function. A more useful circuit enables the UP/DOWN function to be controlled by a separate control line. The line could be controlled by a switch or even from a computer using a software command.

You have seen how reversing the bits causes the count sequence to reverse. This occurs for any nontruncated binary counting sequence, no matter how long. You can take advantage of this idea by either passing the output bits on unchanged or reversing them. The 7486 XOR gate allows you to do this. Each output from the 7493A can be connected to one input of an XOR gate. The other input is connected to an up/down control line. This enables the counter to count either up or down; but when the sequence is reversed, it causes the output number to change immediately, a disadvantage to this technique.

Design the circuit using the 7493A as a 4-bit (0 to 15) counter. Add the up/down control using an SPST switch to select between up or down counting. Draw your circuit in the report and test its operation. Be sure to show all connections to the 7493A, including the reset and clock lines. Summarize your findings in your report.

Multisim Troubleshooting (Optional)

The companion website for this manual has Multisim 11 and 12 files. Download the file named Exp-21nf and the worksheet Exp-21ws. Open the file named Exp-21nf. A four-channel oscilloscope is set up for you already. Notice that it is easier to see the glitches in the computer simulation than in the lab because they are so fast. Open Exp-21nf to analyze the circuit. Complete the worksheet and attach it to the report.

Report for Experiment 21

Name: _____ Date: _____ Class: _____

Objectives:

☐ Build and analyze asynchronous up and down counters.
☐ Change the modulus of a counter.
☐ Use an IC counter and truncate its count sequence.

Data and Observations:

Waveforms from Step 1:

Clock:

A:

B:

PLOT 1

Is this an up counter or a down counter? _____

Waveforms from Step 2:

Clock:

A:

B:

PLOT 2

Is this an up counter or a down counter? _____

Waveforms from Step 3:

Clock:

A:

B:

PLOT 3

Is this an up counter or a down counter? _____

Waveforms from Step 4:

Clock:

A:

B:

PLOT 4

Is this an up counter or a down counter? _____

Waveforms from Step 5:

Clock:

A:

B:

PLOT 5

What is the count sequence for this counter? _____

Waveforms from Step 6:

Clock:

A:

B:

$\overline{\text{State 0}}$
(decoded output):

PLOT 6

Waveforms from Step 7:

Q_A:

Q_B:

Q_C:

Q_D:

PLOT 7

Waveforms from Step 8:

Q_A:

Q_B:

Q_C:

Q_D:

PLOT 8

What is the count sequence of the counter? _____

Results and Conclusion:

Further Investigation Results:

Evaluation and Review Questions

1. Figure 21–8 shows a digital oscilloscope display for the circuit in 21–2, but with a different clock frequency. The upper signal is *B* on CH1. The lower signal is *A* on CH2.
   a. What is the clock frequency?

   b. Why is it best to trigger the scope on CH1?

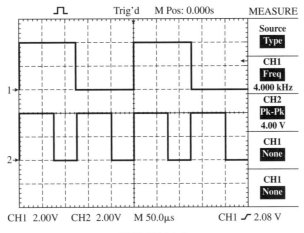

FIGURE 21–8

2. a. Explain how the count sequence of a ripple counter can be truncated.

b. Why does the procedure produce a glitch?

3. Suppose that the counter in Figure 21–2 has both LEDs on all the time. The clock is checked and found to be present. What possible faults would cause this condition?

4. Why shouldn't you trigger your oscilloscope from the clock when determining the time relationship of the outputs of a counter?

5. a. Draw the circuit for a 7493A configured as a modulus-9 counter.

b. Sketch the waveforms you would see from the circuit.

Q_A

Q_B

Q_C

Q_D

6. Assume the 7493A in Figure 21–7 were replaced with a 7492A and wired the same way. Referring to the count sequence shown on the data sheet, determine the modulus and count sequence of the circuit.*

---

*See Appendix A.

Experiment 22
Analysis of Synchronous Counters with Decoding

Objectives

After completing this experiment, you will be able to

☐ Analyze the count sequence of synchronous counters using a tabulation method.

☐ Construct and analyze a synchronous counter with decoding. Draw the state diagram.

☐ Use an oscilloscope to measure the time relationship between the flip-flops and the decoded outputs.

☐ Explain the concept of partial decoding.

Materials Needed

Two 74LS76A dual J-K flip-flops
7400 quad NAND gates
Two SPST N.O. pushbuttons
Four LEDs
Resistors: four 330 Ω, two 1.0 kΩ

For Further Investigation:
One MAN-72 seven-segment display
Seven 470 Ω resistors

Summary of Theory

Synchronous counters have all clock lines tied to a common clock, causing all flip-flops to change at the same time. For this reason, the time from the clock pulse until the next count transition is much faster than in a ripple counter. This greater speed reduces the problem of glitches (short, unwanted signals due to nonsynchronous transitions) in the decoded outputs. However, glitches are not always eliminated because stages with slightly different propagation delays can still have short intermediate states. One way to eliminate glitches is to choose a Gray code count sequence (only one flip-flop transition per clock pulse).

Decoding is the "detecting" of a specific number. A counter with full decoding has a separate output for each state in its sequence. The decoded output can be used to implement some logic that performs a task. The decoded outputs are also useful for developing counters with irregular count sequences. This experiment will also introduce you to *partial* decoding, a technique that allows you to decode the output with less than all of the bits.

A number of MSI counters are available with features on one chip, such as synchronous and asynchronous preset or clear, up/down counting, parallel loading, display drivers, and so on. If it is possible to use an MSI counter for an application, this choice is generally the most economical. If it is not possible, then you must design the counter to meet the requirement. In this experiment you will analyze already designed synchronous counters step by step. In the next experiment, you will design a counter to meet a specific requirement.

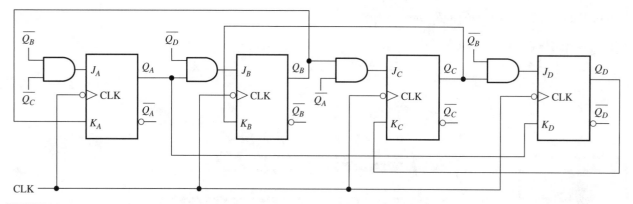

FIGURE 22–1
Synchronous counter with irregular sequence used for half-stepping a stepper motor.

Analysis of Synchronous Counters

The method for analyzing a synchronous counter is a systematic tabulation technique. This method is illustrated for the counter shown in Figure 22–1. Begin by setting up Table 22–1(a). The table lists the outputs and inputs for each flip-flop in the counter.

Step 1. Write the equations for the J and K inputs of each counter using the schematic.

Step 2. Assume counter is in some state; in this example, it is arbitrarily placed in state 0000_2.

Step 3. Complete first row by determining each J and K input. The equations and the inputs 0000_2 are used to compute the binary value of J and K.

Step 4. Use the J-K truth table to determine the next state of each flip-flop. In this example, $J_D = K_D = 0$ means Q_D will not change; Q_C and Q_B also will not change, but $J_A = 1$, $K_A = 0$ means that $Q_A = 1$ after the next clock pulse. Write the next state under the present state that was originally assumed.

Step 5. Continue until all possible inputs are accounted for. This is done in Table 22–1(b). The sequence can now be shown on a state diagram, as in Figure 22–2.

The analysis continues along these lines until all possible (2^N) states have been taken into account, including states that are not in the main count sequence. The completed table is shown as Table 22–1(b). Using the information from the table, the complete state diagram can then be drawn as illustrated in Figure 22–2. This completely describes the operation of the counter. This particular counter has an interesting and somewhat unusual application. It

is used to develop the proper sequence of signals necessary to half-step a stepper motor.

Procedure

Analysis of Synchronous Counters

1. Examine the counter shown in Figure 22–3. Since there are two flip-flops, there are four possible output states. Analyze the sequence by the method illustrated in the Summary of Theory section. Complete Table 22–2 in the report. From the table, draw the predicted state diagram in the space provided in the report.

2. Build the circuit. Use a TTL-level pulse generator at 10 kHz for the clock signal. The NAND gates serve as state decoders with an active-LOW output for each state. To avoid confusion, the lines from the counter to the decoders are not shown on the schematic. If you have a logic analyzer available, look at the outputs of the two flip-flops and the four decoders at the same time. If you do not have a logic analyzer, you can establish the relative time between signals using a two-channel oscilloscope. The following procedure will guide you:

a. Set up the scope to trigger on channel 1 with the Q_B signal (the slowest signal) viewed on that channel. If you are using an analog scope, do not use composite or vertical-mode triggering.
b. View the pulse generator (clock) on channel 2. Adjust the frequency or SEC/DIV control so that each clock pulse coincides with a major division on the horizontal axis.
c. Do not change the triggering or the Q_B signal on the triggering channel. Probe the circuit with

TABLE 22–1
Analysis of synchronous counter shown in Figure 22–1.

	Outputs				Inputs							
	Q_D	Q_C	Q_B	Q_A	$J_D = \overline{Q}_B \cdot Q_C$	$K_D = Q_A$	$J_C = \overline{Q}_A \cdot Q_B$	$K_C = Q_D$	$J_B = Q_A \cdot \overline{Q}_D$	$K_B = Q_C$	$J_A = \overline{Q}_B \cdot \overline{Q}_C$	$K_A = Q_B$
Step 2	0	0	0	0	0	0	0	0	0	0	1	0
	0	0	0	1								
Step 4												

(a) Steps in filling out the table

	Outputs				Inputs							
	Q_D	Q_C	Q_B	Q_A	$J_D = \overline{Q}_B \cdot Q_C$	$K_D = Q_A$	$J_C = \overline{Q}_A \cdot Q_B$	$K_C = Q_D$	$J_B = Q_A \cdot \overline{Q}_D$	$K_B = Q_C$	$J_A = \overline{Q}_B \cdot \overline{Q}_C$	$K_A = Q_B$
Main sequence	0	0	0	0	0	0	0	0	0	0	1	0
	0	0	0	1	0	1	0	0	1	0	1	0
	0	0	1	1	0	1	0	0	1	0	0	1
	0	0	1	0	0	0	1	0	0	0	0	1
	0	1	1	0	0	0	1	0	0	1	0	1
	0	1	0	0	1	0	0	0	0	1	0	0
	1	1	0	0	1	0	0	1	0	1	0	0
	1	0	0	0	0	0	0	1	0	0	1	0
	1	0	0	1	0	1	0	1	0	0	1	0
	0	0	0	1	At this step, a repeated pattern is noted.							
Account for all other states	1	1	0	1	1	1	0	1	0	1	0	0
	0	0	0	1	Returns to main sequence							
	0	1	0	1	1	1	0	0	1	1	0	0
	1	1	1	1	0	1	0	1	0	1	0	1
	0	0	0	0	Returns to previously tested state (0000)							
	0	1	1	1	0	1	0	0	1	1	0	1
	0	1	0	1	Returns to previously tested state (0101)							
	1	0	1	0	0	0	1	1	0	0	0	1
	1	1	1	0	0	0	1	1	0	1	0	1
	1	0	0	0	Returns to main sequence							
	1	0	1	1	0	1	0	1	0	0	0	1
	0	0	1	0	Returns to main sequence							

(b) Completed table

channel 2. The observed signals will be in the proper relationships to the Q_B signal.

Now on Plot 1 in the report, sketch the outputs of the flip-flops and decoders in the proper time relationship to each other.

3. Looking at the waveforms you have drawn, check that your predicted state diagram is correct.

As an extra check, you can slow the clock to 1 Hz and verify the sequence with the LEDs.

4. Assume that a failure has occurred in the circuit. The wire from the Q_B output to K_A has become open. What effect does this open have on the output? Look at the signals and determine the new state diagram.

FIGURE 22–2
Analysis of synchronous counter shown
in Figure 22–1 gives this state diagram.
Analysis procedure is shown in Table
22–1(b).

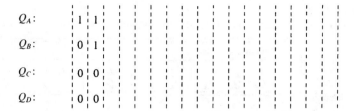

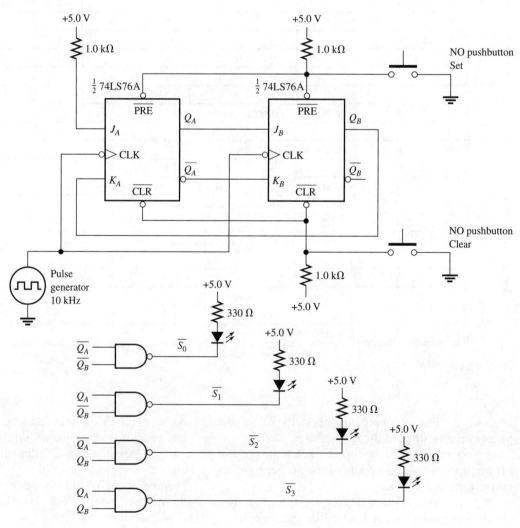

FIGURE 22–3
Synchronous counter with state decoding.

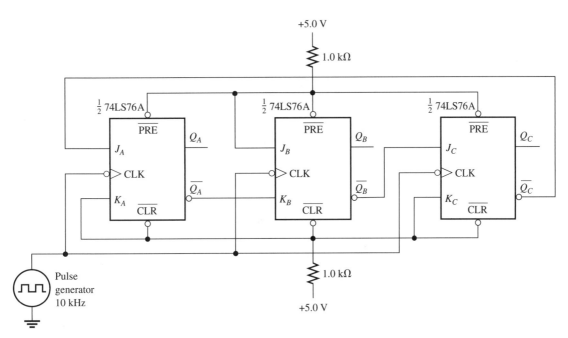

+5.0 V

$1.0 \text{ k}\Omega$

$\frac{1}{2}$ 74LS76A

$\overline{PRE}$

J_A

CLK

K_A $\overline{CLR}$

Q_A

$\overline{Q_A}$

$\frac{1}{2}$ 74LS76A

$\overline{PRE}$

J_B

CLK

K_B $\overline{CLR}$

Q_B

$\overline{Q_B}$

$\frac{1}{2}$ 74LS76A

$\overline{PRE}$

J_C

CLK

K_C $\overline{CLR}$

Q_C

$\overline{Q_C}$

Pulse generator 10 kHz

$1.0 \text{ k}\Omega$

+5.0 V

FIGURE 22–4

Draw the predicted state diagram in your report. Test your prediction by opening the K_A input and observing the result. You can put the counter into state 0 by pressing the clear pushbutton and into state 3 by pressing the set pushbutton.

5. Modify the circuit by adding another flip-flop and changing the inputs to J_A, K_A, and J_B, as shown in Figure 22–4. Leave the 7400 decoder circuit, but remove the set and clear switches. The decoder circuit will form an example of *partial* decoding—a technique frequently employed in computers.

6. Analyze the counter by completing Table 22–3 in the report. Account for all possible states, including unused states. If you correctly account for the unused states, you will see that all unused states return to state 2. Draw the state diagram.

7. Set the pulse generator for 1 Hz and observe the LEDs connected to the state decoders. Notice that state 4 is *not* in the main sequence but state 0 *is* in the main sequence of the counter. This means that every time the state 0 LED turns ON, the counter is actually in state 0. This is an example of partial decoding; the MSB was not connected to the decoder, yet there is no ambiguity for state 0 because state 4 is not possible. Likewise, there is no ambiguity for state 0 or state 7,

but partial decoding is not adequate to uniquely define states 2 and 6.

For Further Investigation

A unique circuit is shown in Figure 22–5. The output is connected in a rather unusual way directly to a seven-segment display. You are challenged to figure out the sequence of letters that will be on the display. Here is your only clue: It is an English word that has something to do with detective work. (You get the other clue when you build the circuit.) If you give up, build the circuit and find the answer.

Multisim Troubleshooting (Optional)

The companion website for this manual has Multisim 11 and 12 files. Download the file named Exp-22nf and the worksheet Exp-22ws. Open the file named Exp-22nf. You may want to view the logic analyzer outputs first and test the C (Clear) and S (Set) switches to verify the circuit performs the same way as your laboratory circuit. Open Exp-22nf to analyze the circuit. Complete the worksheet and attach it to the report.

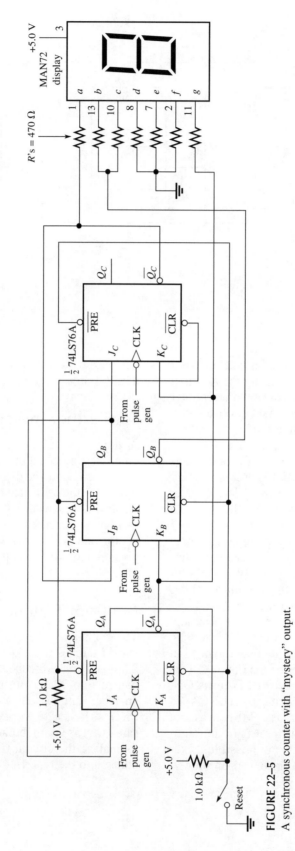

FIGURE 22–5
A synchronous counter with "mystery" output.

168

Report for Experiment 22

Name: _____ Date: _____ Class: _____

Objectives:

☐ Analyze the count sequence of synchronous counters using a tabulation method.
☐ Construct and analyze a synchronous counter with decoding. Draw the state diagram.
☐ Use an oscilloscope to measure the time relationship between the flip-flops and the decoded outputs.
☐ Explain the concept of partial decoding.

Data and Observations:

TABLE 22–2
Analysis of synchronous counter shown in Figure 22–3.

Outputs		Inputs			
Q_B	Q_A	$J_B =$	$K_B =$	$J_A =$	$K_A =$

State diagram:

Q_A:
Q_B:
$\overline{S_0}$:
$\overline{S_1}$:
$\overline{S_2}$:
$\overline{S_3}$:

PLOT 1

169

Step 4. State diagram:

TABLE 22–3
Analysis of synchronous counter shown in Figure 22–4.

Outputs	Inputs					
Q_C $\quad Q_B$ $\quad Q_A$	$J_C =$	$K_C =$	$J_B =$	$K_B =$	$J_A =$	$K_A =$

Step 6. State diagram:

Results and Conclusion:

Further Investigation Results:

Evaluation and Review Questions

1. The counter used for half-stepping a stepper motor in the example of Figure 22–1 has a state diagram sequence that is shown in Figure 22–2. Beginning with state 1, sketch the Q_D, Q_C, Q_B, and Q_A outputs. (*Hint:* An easy way to start is to write the binary number vertically where the waveforms begin. This procedure is started as an example.)

Q_A:　1　1

Q_B:　0　1

Q_C:　0　0

Q_D:　0　0

2. Determine the sequence of the counter shown in Figure 22–6 and draw the state diagram.

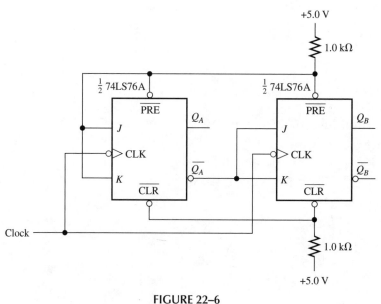

FIGURE 22–6

3. How could you incorporate full decoding into the counter circuit shown in Figure 22–4?

4. Explain the changes you would make to the circuit in Figure 22–4 in order to add a pushbutton that resets the counter into state 2.

5. Assume a problem exists with the counter shown in Figure 22–3. The counter is "locked-up" in state 3. What are two faults that can account for this problem?

6. Assume the synchronous counter in Figure 22–1 is "locked-up" in state 9. A quick check of power, ground, and clock indicate they are all O.K. Which flip-flop is the likely cause of the problem? Why?

Experiment 23
Design of Synchronous Counters

Objectives

After completing this experiment, you will be able to

☐ Design a synchronous counter with up to 16 states in any selected order.

☐ Construct and test the counter. Determine the state diagram of the counter.

Materials Needed

Two 74LS76A dual J-K flip-flops
7408 quad AND gate or other SSI IC determined by student

For Further Investigation:
   74LS139A dual 2-to-4 line decoder
   Six LEDs

Summary of Theory

The design of a synchronous counter begins with a description of the state diagram that specifies the required sequence. All states in the main sequence should be shown; states that are not in the main sequence should be shown only if the design requires these unused states to return to the main sequence in a specified way. If the sequence can be obtained from an already existing IC, this is almost always more economical and simpler than designing a special sequence.

From the state diagram, a next-state table is constructed. This procedure is illustrated with the example in Figure 23–1 for a simple counter and again in Figure 23–3 for a more complicated design. Notice in Figure 23–1 that the next state table is just another way of showing the information contained in the state diagram. The advantage of the table is that the changes made by each flip-flop going from one state to the next state are clearly seen.

The third step is to observe the transitions (changes) in each state. The required logic to force these changes will be mapped onto a Karnaugh map. In this case, the Karnaugh map takes on a different meaning than it did in combinational logic but it is read the same way.* Each square on the map represents a state of the counter. In effect, the counter sequence is just moving from square to square on the Karnaugh map at each clock pulse. To find the logic that will force the necessary change in the flip-flop outputs, look at the transition table for the J-K flip-flop, shown as Table 23–1. Notice that all possible output *transitions* are listed first; then the inputs that cause these changes are given. The transition table contains a number of X's (don't cares) because of the versatility of the J-K flip-flop, as explained in the text. The data from the transition table are entered onto the Karnaugh maps as illustrated.

---

*This type of Karnaugh map may be more properly termed a Karnaugh state map.

Assume you need to design a counter that counts 0–1–3–2 and stays in state 2 until a reset button is pressed. Two flip-flops are required. Let Q_B = MSB and Q_A = LSB. Use a J-K flip-flop.

Step 1: Draw a state diagram.

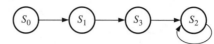

Step 2: Draw next-state table.

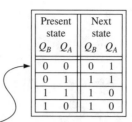

Present state		Next state	
Q_B	Q_A	Q_B	Q_A
0	0	0	1
0	1	1	1
1	1	1	0
1	0	1	0

Step 3: Determine inputs required for each flip-flop.
(a) Read present state 00 on next-state table.
(b) Note that Q_B does not change $0 \rightarrow 0$ (present to next state) and Q_A changes from $0 \rightarrow 1$.
(c) Read the required inputs to cause these results from transition Table 20-1.
(d) Map each input from transition table onto Karnaugh map.

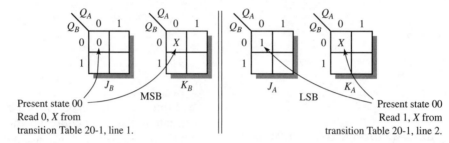

Present state 00
Read 0, X from
transition Table 20-1, line 1.

MSB

LSB

Present state 00
Read 1, X from
transition Table 20-1, line 2.

(e) Complete maps.

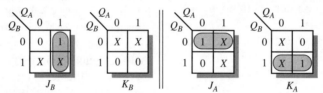

(f) Read minimum logic from each map.

$$J_B = Q_A, \qquad K_B = 0, \qquad J_A = \overline{Q_B}, \qquad K_A = Q_B$$

Step 4: Draw circuit and check.

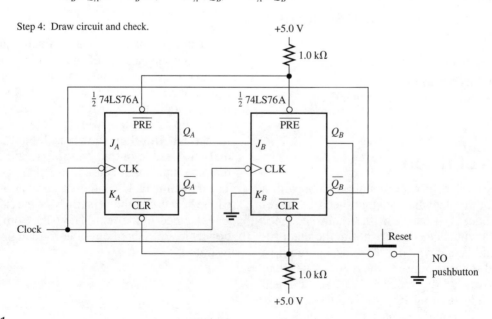

FIGURE 23–1

TABLE 23–1

Transition table for J-K flip-flop.

Output Transitions		Inputs	
Q_N	Q_{N+1}	J_N	K_N
0 $\rightarrow$ 0		0	X
0 $\rightarrow$ 1		1	X
1 $\rightarrow$ 0		X	1
1 $\rightarrow$ 1		X	0

Q_N = output before clock

Q_{N+1} = output after clock

J_N, K_N = inputs required to cause transition

X = don't care

When the maps are completed, the logic can be read from the map. This logic is then used to set up the circuit as shown in Step 4 of Figure 23–1. It is a good idea to check the design by verifying that the count sequence is correct and that there are no lock-up states. (A lock-up state is one that does not return to the main sequence of the counter.) The design check can be done by completing a table such as Table 22–1 in the last experiment.

The design procedure just described can be extended to more complicated designs. In Experiment 22 a counter was shown (Figure 22–1) that generates the required waveforms for half-stepping a stepper motor. This counter produces the state sequence shown in Figure 23–2(a). This sequence can be drawn as a series of four waveforms required by the stepper motor as shown in Figure 23–2(b).

The design method described here is not the only way to obtain the desired sequence, but it does lead to a fairly straightforward design. Figure 23–3 illustrates the detailed procedure for designing this circuit. Note that only the main sequence is shown in the state diagram and on the next-state table. The

reason for this is that the unused states will show up as extra "don't cares" in the logic, making the design simpler. All unused states are entered on the maps as "don't cares." After reading the logic equations for the inputs to each flip-flop, the design is checked for lock-up problems. Corrections are made to prevent lock up by examining the "don't-care" logic and changing it if required. The maps for the *A* and *B* flip-flops are not shown in Figure 23–3, but left as a student exercise in Question 1 of the Evaluation and Review Questions.

As you can see in Figure 23–3, the steps for the more complicated counter are basically the same as those used in Figure 23–1. The unused states allow the counter to be designed with a minimum of additional logic. The completed design is shown in Figure 22–1; it is the same circuit that was analyzed in Experiment 22.

Procedure

1. A Gray code synchronous counter is often used in state machine design. This problem requires a six-state Gray code counter. The usual Gray code sequence is not used because the sixth state would not be "Gray" when the counter returns to zero. Instead, the sequence shown in Figure 23–4 is required. There are two unused states: state 5 and state 7. For the initial design, these states are not shown. Complete the next-state table in the report for the main sequence shown here.

2. Using the transition table for the J-K flip-flop, complete the Karnaugh maps shown in the report. The J-K transition table (Table 23–1) is repeated in the report for convenience.

3. Read the required logic expressions from each map that you completed in step 2. Check that the unused states return to the main sequence. If they do not, modify the design to assure that they do return. Then, construct and test your circuit. You can

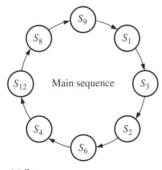

(a) State sequence

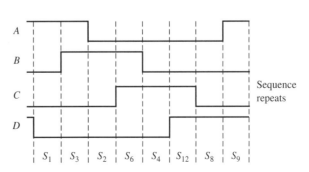

Sequence repeats

(b) Waveform representation of state sequence

FIGURE 23–2

Step 1: Draw the required state diagram. (Note that only the main sequence is shown as the unused states are not important in this problem.)

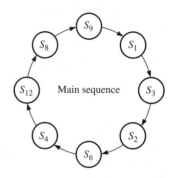

Step 2: Draw the next-state table. Four flip-flops are required because of the number of bits used in the sequence.

Present state				Next state			
Q_D	Q_C	Q_B	Q_A	Q_D	Q_C	Q_B	Q_A
0	0	0	1	0	0	1	1
0	0	1	1	0	0	1	0
0	0	1	0	0	1	1	0
0	1	1	0	0	1	0	0
0	1	0	0	1	1	0	0
1	1	0	0	1	0	0	0
1	0	0	0	1	0	0	1
1	0	0	1	0	0	0	1

Step 3: Using the next-state and transition tables, draw the Karnaugh maps for each flip-flop. For example, in state 1, note that Q_D and Q_C do not change in going to the next state. The transition is $0 \rightarrow 0$. From the transition table, a $0 \rightarrow 0$ transition requires $J = 0$ and $K = X$. These values are entered onto the maps for the D and C counters in the square that represents state 1. Unused states are mapped as Xs. Only the D and C maps are shown in this example.

(Note: Q_B and Q_A are positioned to make the map below easier to read.)

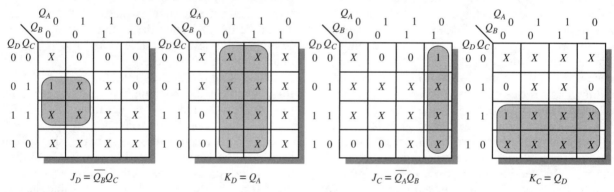

FIGURE 23–3

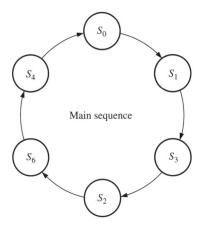

FIGURE 23–4
Required sequence for Gray code counter.

check the state sequence with an oscilloscope or a logic analyzer. Summarize the results of your test in your report.

For Further Investigation

A decoded output is needed for the counter you designed. Unfortunately, the only decoder IC that engineering has available for decoding is a 2-line to 4-line 74LS139A decoder! Show how you could connect this IC to obtain full decoding of the output. Then construct the circuit and put a separate LED on each output so that only one LED lights as the counter goes around. (*Hint:* Consider how you could use the enable inputs of the 74LS139A.)

Report for Experiment 23

Name: _____ Date: _____ Class: _____

Objectives:

☐ Design a synchronous counter with up to 16 states in any selected order.
☐ Construct and test the counter. Determine the state diagram of the counter.

Data and Observations:

Present State			Next State		
Q_C	Q_B	Q_A	Q_C	Q_B	Q_A
0	0	0			
0	0	1			
0	1	1			
0	1	0			
1	1	0			
1	0	0			

TABLE 23–1
Transition table for J-K flip-flop (repeated for reference).

Output Transitions		Inputs	
Q_N	Q_{N+1}	J_N	K_N
0	0	0	X
0	1	1	X
1	0	X	1
1	1	X	0

Q_N = output before clock
Q_{N+1} = output after clock
J_N, K_N = inputs required to cause transition
X = don't care

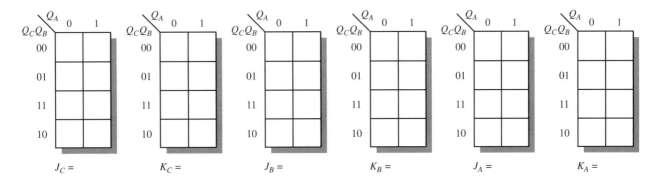

$J_C =$ $K_C =$ $J_B =$ $K_B =$ $J_A =$ $K_A =$

Circuit Design:

179

Results and Conclusion:

Further Investigation Results:

Evaluation and Review Questions

1. Complete the design of the sequential counter in Figure 23–3 by constructing Karnaugh maps for the *B* and *A* flip-flops. Read the maps. As a check, you can compare your result with the circuit drawn in Figure 22–1.

2. Describe the logic necessary to add a seven-segment display to the circuit you designed in this experiment to enable the display to show the state of the counter.

3. Assume you wanted to make the sequential circuit you designed in this experiment start in state 6 if a reset pushbutton is pressed. Describe how you would modify the circuit to incorporate this feature.

4. Assume you wanted to change the circuit from this experiment to be able to reverse the sequence. How would you go about this?

5. Assume you wanted to trigger a one-shot (74121) whenever the circuit you designed went into state 2 or state 4. Explain how you could do this.

6. a. Draw the transition table for a D flip-flop. Start by showing all possible output transitions (as in the J-K case) and consider what input must be placed on D in order to force the transition.

   b. Why is the J-K flip-flop more versatile for designing synchronous counters with irregular sequences?

Experiment 24
The Traffic Signal Controller

Objectives

After completing this experiment, you will be able to
- [ ] Complete the design of a sequential counter that is controlled by input variables.
- [ ] Construct and test the circuit from the first objective.

Materials Needed

7408 quad AND gate
7474 dual D flip-flop
74121 one-shot
74LS153 dual data selector
One 150 μF capacitor
Two LEDs
Resistors: two 330 Ω, six 1.0 kΩ, one to be determined by student

Summary of Theory

A synchronous counter forms the heart of many small digital systems. The traffic signal controller introduced in Experiments 13 and 17 uses a small synchronous counter to represent each of the four possible "states" that the output can take. The block diagram of the system was given in Figure 13–4. Unlike the counters in Experiment 23, the state of the counter in the traffic signal controller is determined by three input variables and two state variables.

When certain conditions of these variables are met, the counter advances to the next state. The three input variables are defined as follows:

Vehicle on side street = V_s

25 s timer (long timer) is on = T_L

4 s timer (short timer) is on = T_S

The use of complemented variables indicates the opposite conditions. A state diagram, introduced in the text, is repeated in Figure 24–1 for reference. Based on this state diagram, the sequential operation is described as follows:

1st state: The counter shows the Gray code 00, representing main-green, side-red. It will stay in the first state if the long timer is on *or* if there is no vehicle on the side street ($T_L + \overline{V}_s$). It will go to the second state if the long timer is off *and* there is a vehicle on the side street ($\overline{T}_L V_s$).

2nd state: The counter shows 01, representing main-yellow, side-red. It will stay in the second state if the short timer is on (T_S). It will go to the third state if the short timer is off ($\overline{T}_S$).

3rd state: The counter shows 11, representing main-red, side-green. It will stay in the third state if the long timer is on *and* there is a vehicle on the side street ($T_L V_s$). It will go to the fourth state if the long timer is off *or* there is no vehicle on the side street ($\overline{T}_L + \overline{V}_s$).

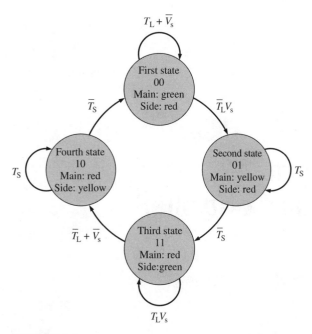

FIGURE 24–1
State diagram.

4th state: The counter shows 10, representing main-red, side-yellow. It will stay in the fourth state if the short timer is on (T_S). It will go to first state if the short timer is off ($\overline{T}_S$).

The block diagram in Figure 24–2 further defines the sequential logic. The input logic block consists of two data selectors to route the three input variables (V_s, T_L, and T_S) to the flip-flops. This is shown in more detail in the partially completed schematic shown in the report. The data selectors (DS-0 and DS-1) are the 74LS153 chips. The line selected (C_0 through C_3) is determined by the present state (because the flip-flop outputs are connected to the select inputs). Notice the similarity of

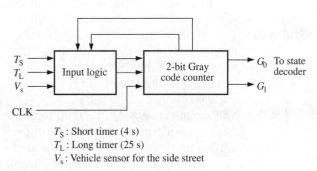

T_S: Short timer (4 s)
T_L: Long timer (25 s)
V_s: Vehicle sensor for the side street

FIGURE 24–2
Block diagram of the sequential logic.

this idea to that shown in Experiment 12 for implementing combinational logic with a MUX.

In this experiment, you will design and construct a portion of the system. To make our simulation more realistic, it will be helpful to have the short timer working. The short timer was constructed from a 74121 one-shot in Experiment 17. The same IC will be used but the trigger connections will be different since the trigger logic is not available for this part of the experiment. From the next-state table, it can be seen that the short timer should be ON in the second and fourth states (Gray codes 01 and 10). Triggering will be set up so that the short timer will trigger on the trailing edge of the clock, causing it to start in any state. This won't matter for this experiment because it is tested only to move into states 00 and 11. The trailing edge is used for triggering the short timer because the outputs change on the leading edge. This avoids a "race" condition where the clock and states change together. The triggering for this experiment is shown in Figure 24–3.

The present state–next state table with input conditions is shown in Table 24–1 of the report. Each pair of lines in Table 24–1 represents the two

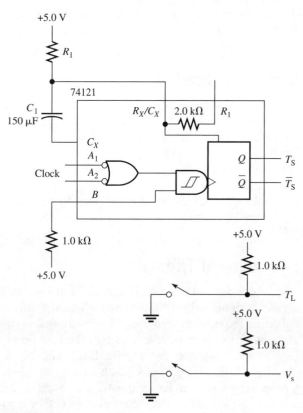

FIGURE 24–3

possible states that the counter could assume. For example, on the first pair of lines, the counter is in the first state (Gray code 00) and could either stay in the first state (Gray code 00) or go to the second state (Gray code 01), depending on the long timer and vehicle sensor inputs. Notice that in the first state (Gray code 00), the next state for Q_1 requires that it remain a 0 *no matter what the inputs do,* so a 0 is entered for the product term for data selector-1 (DS-1). On the other hand, Q_0 will be a 1 *only if* the long timer is LOW (timed out) *and* the vehicle sensor is HIGH (vehicle waiting). Thus, the input product term for DS-0 is $\overline{T}_L V_s$. As one more example, notice that in the second state (Gray code 01), the next state for Q_0 will be a 1 *no matter what the inputs do,* so a 1 is entered in the table.

Procedure

1. Review the Summary of Theory to be sure you understand the idea for the circuit. The present state–next state table with input conditions is shown in Table 24–1 of the report. Three of the inputs for the data selectors are completed as an example. Decide what to do with the remaining inputs and complete the remaining five inputs in Table 24–1.

2. From the inputs determined in Step 1, complete the schematic shown in Figure 24–4 of the report. Show the inputs to each data selector and the enable. Note that the select lines of the data selectors are connected to the outputs of the flip-flops.

3. To simulate the inputs, you *could* construct the one-shots and the oscillator from Experiment 17. However, to save time and board space, construct only the short timer shown in Figure 24–3. You will need to compute the value of R_1 in order to make a 4 s timer. Notice that the triggering of the short timer is different than in the full system for reasons of simplicity. The long timer and vehicle sensor are made from SPST switches as shown in Figure 24–3. A "NOTed" variable, such as $\overline{T}_L$, is asserted when the switch is closed.

4. On the same protoboard as the short timer and the switches representing the long timer and vehicle sensor, add the sequential logic in Figure 24–4. The LEDs serve as state indicators. Connect all inputs in accordance with your design in Step 1. Set the pulse generator to 10 kHz.

5. The state diagram (Figure 24–1) will guide you through the inputs required for the sequence in order to test the circuit. Start by opening the long timer and vehicle sensor switches (both HIGH). Place the counter in the first state (Gray code 00) by placing a momentary ground on the $\overline{CLR}$ inputs. In this condition, a vehicle is assumed to be on the side street (because V_s is HIGH) but the long timer has not finished the cycle on the main street. Close the long timer switch. This should immediately cause the circuit to go into the second state (Gray code 01), and the 4 s short timer should take over. While the 4 s timer is on, open the long timer switch again. The circuit should switch by itself (after 4 s) to the third state (Gray code 11).

6. If you successfully arrived in the fourth state (Gray code 10), look at the state diagram and decide what steps need to be taken to return to the first state and remain there. Then, use the switches to move back to the first state. Summarize your results in the Results and Conclusion section.

For Further Investigation

Assume your boss wonders, "Can you simplify the traffic signal controller if you eliminate the short timer and make the clock (pulse generator) operate at a period of 4 s? Does this have any advantages? Also, as a second idea, I noticed you triggered the short timer without using trigger logic, just the clock. Could you also trigger the long timer this way?"

Consider both of these ideas (after all, you shouldn't ignore the boss). Indicate the circuit modifications you would suggest in order to accomplish each. Try putting the first idea into effect by modifying the circuit and testing it again. Then write a short summary to the boss stating what you think of her idea.

Report for Experiment 24

Name: _____ Date: _____ Class: _____

Objectives:

☐ Complete the design of a sequential counter that is controlled by input variables.
☐ Construct and test the circuit from the first objective.

Data and Observations:

TABLE 24–1

Present State		Next State		Input Conditions	Input Product Term for Data Selector-1	Input Product Term for Data Selector-0
Q_1	Q_0	Q_1	Q_0			
0	0	0	0	$T_L + \overline{V}_s$		
0	0	0	1	$\overline{T}_L V_s$	0	$\overline{T}_L V_s$
0	1	0	1	T_S		
0	1	1	1	$\overline{T}_S$		1
1	1	1	1	$T_L V_s$		
1	1	1	0	$\overline{T}_L + \overline{V}_s$		
1	0	1	0	T_S		
1	0	0	0	$\overline{T}_S$		

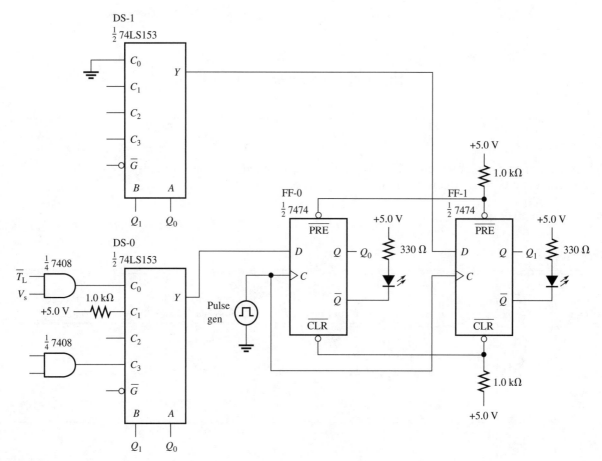

FIGURE 24–4

Results and Conclusion:

Further Investigation Results:

Evaluation and Review Questions

1. Why was Gray code selected for the design of the traffic signal controller?

2. What two conditions are required for the counter to stay in the third state (Gray code 11)?

3. Explain what modifications would be needed to make the traffic signal controller cycle through eight states instead of four states.

4. Suppose you want to build the traffic signal controller using J-K flip-flops instead of D flip-flops. How should the *J* and *K* inputs be connected to allow this change?

5. The *B* input of the 74121 (Figure 24–3) was connected to a HIGH. Explain why.

6. Assume the traffic signal controller is "locked-up" in the first state (Gray code 00). The light never cycles to the side street even when a vehicle is present, causing drivers to become extremely annoyed. You test the vehicle sensor and find that it is HIGH at the input to the 7408 AND gate. Describe the troubleshooting procedure you would use to identify the problem.

Experiment 25
Semiconductor Memories

Objectives

After completing this experiment, you will be able to
- Design and implement the serial to 8-bit parallel converter logic.
- Test and verify your design and results.

Materials Needed

PC with VHDL or Verilog programming software (see Preface).
Optional: Project Board compatible with Quartus II or Project Navigator
RW_RAM memory project
file, available at
    http://www.pearsonhighered.com/floyd.

Summary of Theory

The basic cell in a static RAM (called an SRAM) is a flip flop; it can be set or reset for a write operation or tested without changing its state for a read operation. In addition, the SRAM contains logic gates to control the read and write functions and decoding circuitry as shown in Figure 25-1.read from the memory.

To select the SRAM cell, the address select (AddSel) input is set high. The bit (Bitin) to be stored sets the J-K flip-flop inputs for a set or reset. The Read/Write input is set LOW to enable the input to the J-K flip-flop and a clock pulse clocks stores the data. Setting the Read/Write input to a HIGH outputs the stored bit.

All RAMs are organized into arrays containing the memory cells in rows and columns as illustrated in Figure 25-2. In memories, the number of bits

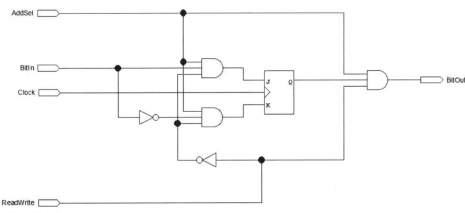

FIGURE 25–1
Basic static RAM cell (SRAM).

191

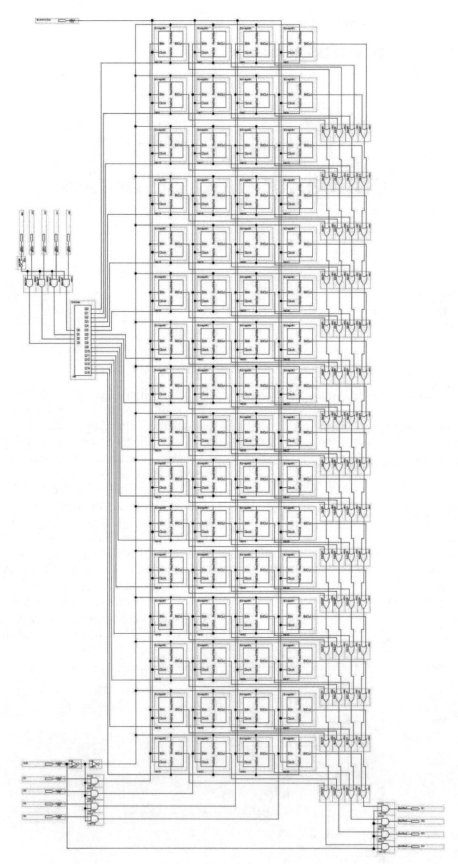

FIGURE 25–2
Logic diagram for a small SRAM R/W memory that was developed for schematic capture.

treated as one entity is considered the word size. A word is the number of bits that are accessed at one time and can vary from as little as 1 bit to as many as 64 bits. For computer applications, a word generally means exactly 16 bits; however, this definition is different in memories and refers to the minimum number of bits written to or read from memory at a time. In Figure 25-2, the word size is 4 bits.

Each word in a Read/Write memory is accessed by a set of address lines, representing the location of the word within the matrix. To access a specific word, the address is decoded; this decoded address is used to select the proper row in memory. Depending on the organization of the matrix, column information may also be required. There are also one or more control lines, which are used to select the read/write option, control the output, or enable the IC. The outputs of memory and other devices are frequently connected to a common set of wires known as a bus. In order for more than one output to be connected to the bus, the output of each device is either a tristate or open-collector type. A tristate device has three logic levels: LOW, HIGH or disconnected (high impedance). Open-collector devices use a single pull-up resistor that is connected to each line. This pull-up resistor keeps the line HIGH unless one of the outputs connected to the line pulls it LOW. Tristate devices do not use a pull-up resistor.

The memory device presented is a basic 4-bit static RAM (SRAM) that will be created by using schematic capture. Each of the memory cells is an individual instance of the Figure 25-1 SRAM cell. An advantage of the SRAM is that it does not require refresh circuitry.

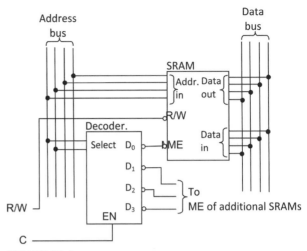

FIGURE 25–3
Expansion of addresses using a decoder.

Four address input bits A0-A3 are enabled by setting input $\overline{ME}$ (memory enable) to a LOW. The input $\overline{ME}$ allows the address inputs to settle before being read to prevent glitches. Glitches are unwanted spikes that can occur if timing is not correct; this is demonstrated in Figure 25-4. Inputs A0-A3 are the inputs to a 4-16 decoded. Although the decoder is also shown as a schematic symbol, the code describing this portion of the device is written in VHDL. Each output of the decoder selects a different 4-bit SRAM array. To store data in the SRAM, the read/write input is set LOW enabling data inputs D1-D4. Setting $R/\overline{W}$ HIGH enables the outputs Q1-Q4 for a read operation.

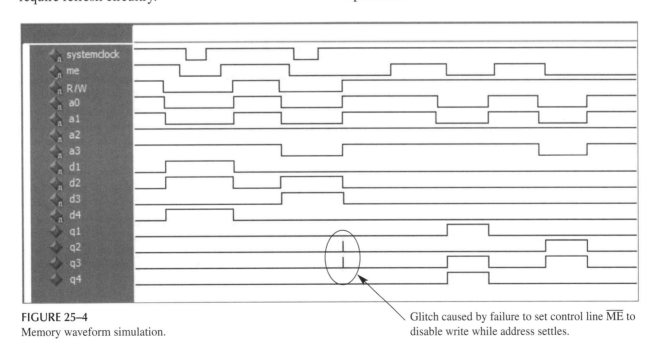

FIGURE 25–4
Memory waveform simulation.

Glitch caused by failure to set control line $\overline{ME}$ to disable write while address settles.

The enable line $\overline{ME}$ allows expansion of memory by allowing other memories to have different external addresses. Figure 25-3 shows how two memory devices can be expanded.

The completed memory is examined through simulation. The systemclock input clocks the bit to be stored into the J-K flip-flop within the memory cell shown in Figure 25-1. The memory enable input $\overline{ME}$ is set HIGH while data and address settle before being set to a LOW for a read or write operation. Notice that a glitch occurs when the $\overline{ME}$ input is left LOW before setting the address. This form of inspection through simulation is critical to a trouble free design.

Procedure

1. Download the RW_RAM memory project provided by the text. All files are available at www.pearsonhighered.com/floyd.

2. Compile your schematic capture design in preparation for project simulation. Refer to the Quartus II schematic capture tutorial.

3. In the space provided in the report, draw the simulation waveform required to store the digits 6, 4, 8, and 9. Include the predicted output in your waveform.

4. Using ModelSim, simulate the waveforms from step 3. Verify that your results match your predictions from step 3. Refer to the ModelSim simulation tutorial.

For Further Investigation

Download the student files for the partially completed security system shown in Figure 25-5.

Refer to the CodeSelection module on the website (http://www.pearsonhighered.com/floyd) that represents the security system presented in Chapter 7 of the textbook. The four-digit combination that is given is written directly in the program code and cannot be changed. Using the static memory presented in this experiment, construct a new CodeSelection module that will allow the user to enter and store a user's four-digit combination. Draw the circuit required and write a short description of your design.

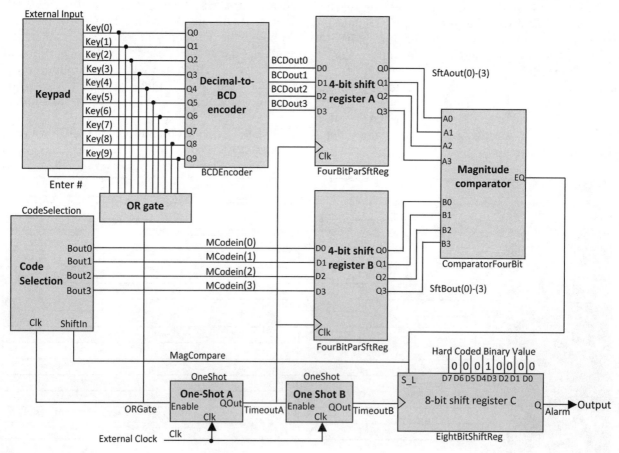

FIGURE 25–5
Security System.

Report for Experiment 25

Name: _____ Date: _____ Class: _____

Data and Observations:

Step 2:

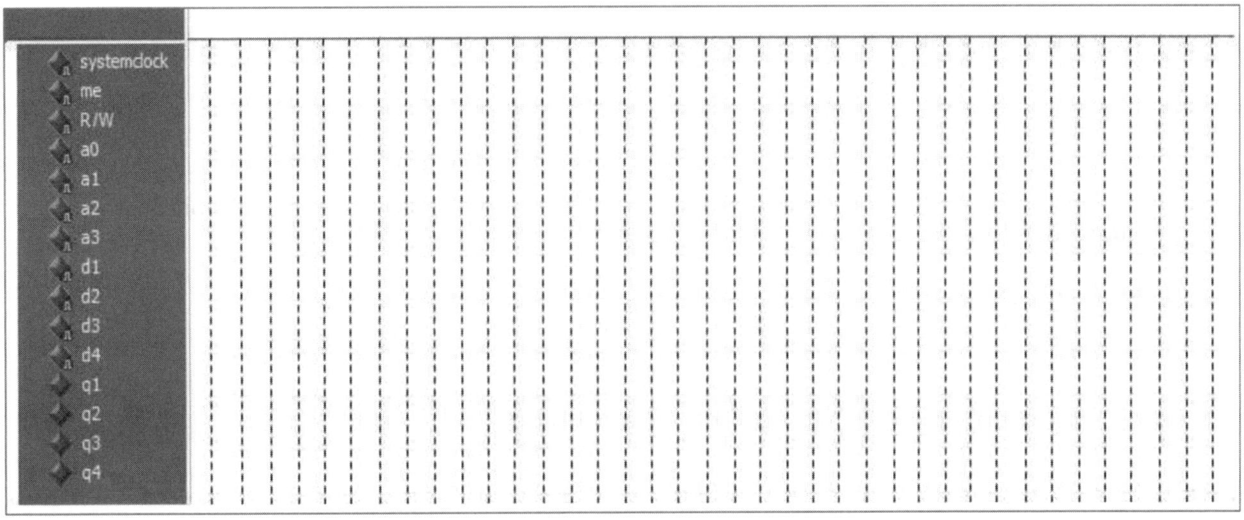

FIGURE 25–6
Memory waveform simulation required to store the digits 6, 4, 8, and 9.

Further Investigation:

Review Questions

1. Explain the operation of the basic SRAM cell shown in Figure 25-1.

2. How can the decoder module in Figure 25-2 be used to decode an 8-bit memory?

3. How is a common-collector device connected to a bus? If you are using a PLD trainer, can your PLD device be connected directly to a bus? Explain your answer.

4. In Figure 25-2 the output of each memory cell is connected to a common output (Q1-Q3) through a series of OR gates.
   a. What is the purpose the OR gates?

   b. What would happen if the OR gates are removed and the output of each memory cell connected directly to the Q1-Q3 outputs?

Experiment 26
Serial-to-Parallel Data Converter

Objectives

After completing this experiment, you will be able to

☐ Design and implement the serial to 8-bit parallel converter logic.

☐ Test and verify your design and results.

Materials Needed

PC with VHDL or Verilog programming software (see Preface). Data file entitled SerialParallelConverter available at http://www.pearsonhighered.com.

Summary of Theory

The serial-to-parallel converter is widely used in systems requiring input from devices connected to USB or serial input ports. In this experiment, a serial-to-parallel converter is used to receive a serial data stream and convert it into eight-bit parallel bytes.

In operation, the eight-bit serial-to-parallel converted needs to know when the message starts as well as when it ends. The data bits are included in a packet of 11 bits as shown in Figure 11-14 of the text. The first bit is a start bit indicating 8 data bits follow. The last two bits can be a parity and stop bit, but for this experiment, two stop bits will be sent to complete the packet. This avoids the extra complexity of a parity generator.

The logic diagram in Figure 26-1 illustrates the basic logic for the serial-to-parallel converter.

The serial-to-parallel converter consists of a control flip-flop that enables the clock generator module. The clock generator module sends the serial input data to the data-input register and increments the divide-by-eight counter. The clock generator and the data input are synchronized so that the 1's and 0's in the data stream are present at the time of the clock generator's leading edge. When the divide-by-eight counter reaches the 9th count, the TC (terminal count) line goes HIGH, signaling that the data should stop clocking into the data input register.

The timing diagram in Figure 26-2 illustrates the operation of the serial-to-parallel converter. An initial HIGH-to-LOW transition of the input serial data stream sets the control flip-flop, which sends the enable signal to the clock generator, which is an astable multivibtrator. The clock generator produces 8-clock pulses that clock data from the input serial data stream to the data-input shift register. It also clocks the CTR DIV 8 counter that sends an end of count signal at count eight to the one-shot enable input. The output of the one-shot clears to the control flip-flop and resets the CTR DIV 8 counter. The output of the one-shot also clocks the data from the data-input shift register to the output shift register. Another HIGH-to-LOW transition of the input data stream starts the process over again. Data is read on a rising clock edge.

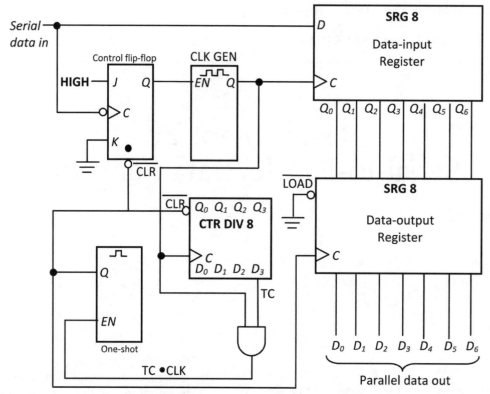

FIGURE 26–1
Basic logic diagram of the serial-to-parallel converter.

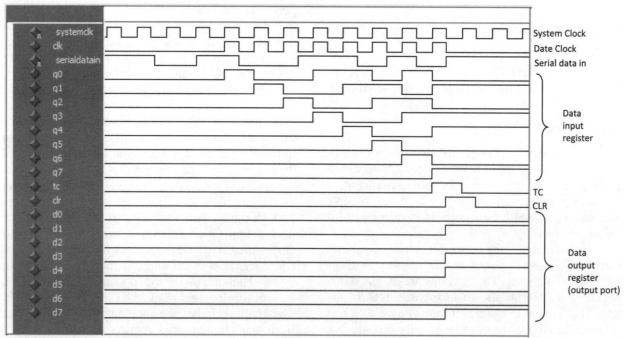

FIGURE 26–2
Timing diagram serial-to-parallel data converter.

198

Procedure

1. Open the SerialParallelConverter project from the student files. From the symbol editor select "Project" and use the block symbols provided to construct the schematic diagram as shown in Figure 26-3.

2. Construct the schematic design shown in Figure 26-4. Refer to the Quartus II schematic capture tutorial.

3. Compile your schematic capture design in preparation for project simulation as described in the Quartus II schematic capture tutorial.

4. Using the ModelSim waveform editor, construct the following waveforms for simulation as shown in Figure 26-5. For this simulation, you will need to create waveforms for the system clock waveform and the serial data input.

5. Verify your simulation results against the serial-to-parallel data converter schematic diagram shown in Figure 26-4.

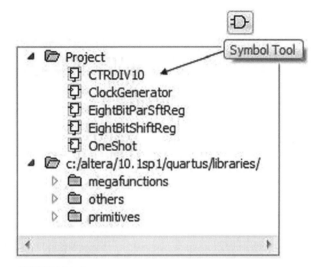

FIGURE 26–3
Work area for schematic capture (step 3).

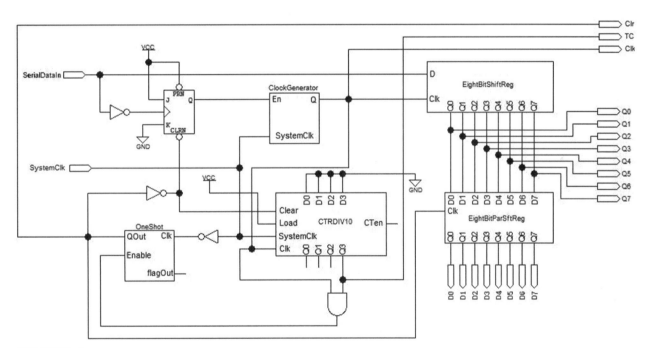

FIGURE 26–4
Serial-to-parallel data converter schematic diagram.

Review Questions

1. Describe how the serial-to-parallel converter can be modified to a sixteen bit converter.

2. Explain why the use of start and stop bits are required in this overall logic of the serial-to-parallel converter.

3. What modifications are required to read serial data on a falling clock edge?

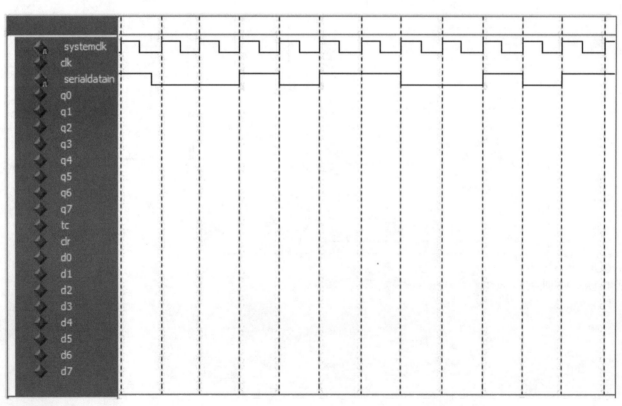

FIGURE 26–5
Simulation input waveforms.

4. How can the circuit be modified to transmit a 4-bit message as shown?

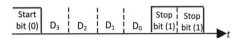

5. Describe the purpose of the control flip-flop.

6. Describe how the one-shot resets the serial to data converter and outputs the current information in the data output register.

Experiment 27
D/A and A/D Conversion

Objectives

After completing this experiment, you will be able to

□ Construct a circuit using a binary counter and a digital-to-analog converter (DAC).

□ Check the calibration of the circuit, measure its resolution in volts/step, and observe its waveforms on an oscilloscope.

□ Modify the circuit into a tracking analog-to-digital converter (ADC), and use it to construct a digital light meter.

Materials Needed

74191 up/down counter
MC1408 digital-to-analog converter*
7447A BCD-to-seven-segment decoder/driver
MAN-72 seven-segment display
LM741 operational amplifier
3906 PNP transistor
CdS photocell (Jameco 120299 or equivalent)
Resistors: seven 330 Ω, two 1.0 kΩ, two 2.0 kΩ, one 5.1 kΩ, one 10 kΩ
One 150 pF capacitor

For Further Investigation:

ADC0804 analog-to-digital converter
Resistors: two 2.2 kΩ, one 1.0 kΩ
One 1.0 kΩ potentiometer
One SPDT switch (wire can be used)

*Available at www.mouser.com.

Summary of Theory

Digital Signal Processing (DSP) has become an important element of many digital designs, especially those that work in *real time*. *Real time* is defined as a time short enough that required processes are completed without affecting the end result. DSP involves three basic steps. These are conversion of an analog signal to digital, processing the signal in a specialized microprocessor, and (usually) conversion of the digital signal back to analog.

The focus of this experiment is the conversion process from analog to digital and back. Conversion from an analog input into a digital quantity is done by an *analog-to-digital converter* (ADC). A *digital-to-analog converter* (DAC) does the reverse—it converts a digital quantity into a voltage or current that is proportional to the input.

A variety of ADCs are available. The selection of a particular one depends on requirements such as conversion speed, resolution, accuracy, and cost. *Resolution* is the number of steps the full-scale signal is divided into. It may be expressed as a percentage of full-scale, as a voltage per step, or simply as the number of bits used in the conversion. *Accuracy,* which is often confused with resolution, is the percentage difference between the actual and the expected output.

A simple analog device, the potentiometer, can be used as a sensor to produce a voltage proportional to the shaft position. Figure 27–1 shows a simplified circuit that illustrates conceptually how the shaft position of the potentiometer can be converted to a digital number. (Only one digit is shown to keep the system simple.) It is drawn with an ADC0804, an 8-bit ADC. The digital output is scaled to a +5.0 V reference—that is, the output is the maximum value when the input is +5.0 V. With a small modification, the output could be scaled for other input voltages. The circuit could be constructed with either the AD673 or the ADC0804, depending on availability. The actual construction of this circuit is left as a Further Investigation.

In the first part of this experiment, you will test an integrated circuit DAC (the MC1408) using a binary up/down counter for a digital input. A reference current of nearly 1.0 mA is set up by a 5.1 kΩ resistor connected to +5.0 V. This current is used to set the full-scale output in the 2.0 kΩ load resistor.

After testing the DAC, you can convert the circuit to a tracking ADC with the addition of a comparator and interface circuity (see Figure 27–3). The binary output from an up/down counter is converted to an analog quantity by the DAC, which is compared to the analog input. This circuit is a little more complicated than that shown in Figure 27–1,

but better illustrates how a tracking ADC works. In principle, any analog input could be used (such as the antenna tracking system input), but we will use a photocell to construct a simplified digital light meter.

Procedure

Multiplying DAC

1. Construct the circuit shown in Figure 27–2. The MC1408 DAC has 8-bit resolution, but we will use only the 4 most significant bits and not bother with a binary-to-BCD decoder. Note how the input pins on the DAC are numbered; the MSB is the lowest number. Set the pulse generator for a TTL-level 1 Hz, and close S_1. Observe the waveforms at the output of the counter. Sketch the observed waveforms in Plot 1 of the report.

2. Open S_1. Observe the waveforms from the counter and draw them in Plot 2.

3. Apply a short to ground on the $\overline{\text{LOAD}}$ input of the counter. This parallel-loads the counter to all 1's. Now check the calibration of the DAC.

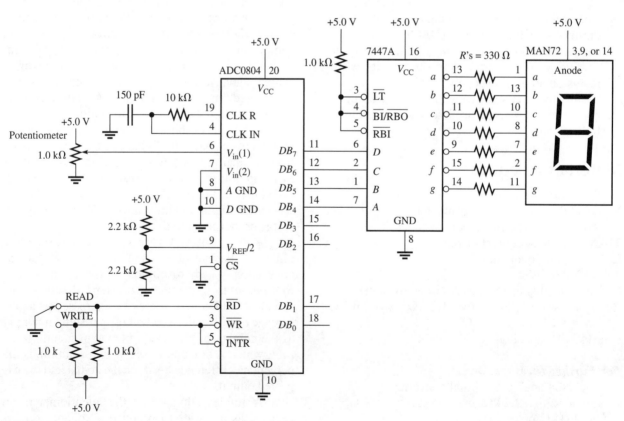

FIGURE 27–1

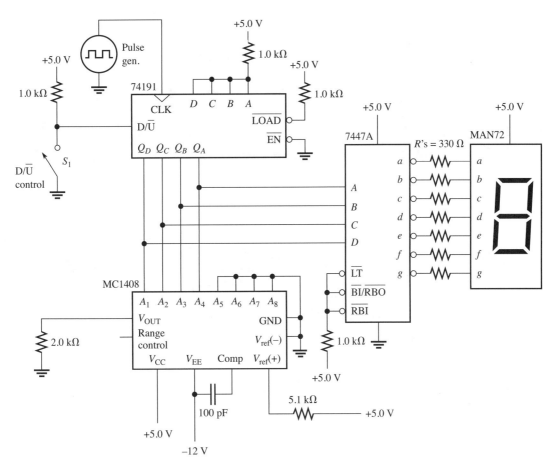

FIGURE 27–2

With the short in place, measure the output voltage on pin 4 of the MC1408. This represents the full-scale output. Determine the volts per step by dividing the full-scale output voltage by 15 (since there are 15 steps present). Record the voltage and the volts per step in Table 27–1 in the report.

4. Disconnect the short to ground from the $\overline{\text{LOAD}}$ input of the counter. Speed the pulse generator up to 1 kHz. With an oscilloscope, observe the analog output (pin 4) from the DAC. On Plot 3 sketch the waveform you see. Label the voltage and time on your sketch.

5. Close S_1 and observe the analog output from the DAC. Sketch the observed waveform on Plot 4.

A Digital Light Meter—A Tracking ADC

6. Modify the circuit into a tracking ADC by adding the photocell, operational amplifier (op-amp), and transistor, as shown in Figure 27–3. The circuit senses light, producing a voltage at the non-inverting input (pin 3) of the op-amp. The input voltage from the photocell is compared to the output

voltage from the DAC and causes the counter to count either up or down. The purpose of the transistor is to change the op-amp output to TTL-compatible levels for the counter.

7. Set the pulse generator for 1 Hz, and note what happens to the count in the seven-segment display as you cover the photocell with your hand. If a constant light is allowed to strike the photocell, the output still oscillates. This oscillating behavior is characteristic of the tracking ADC. Speed up the generator and observe the analog output of the DAC on the oscilloscope. Describe the signal observed on the scope as you cover the photocell.

8. The modification shown in Figure 27–4 is a simple way to remove the oscillating behavior from the seven-segment display. Note that the Q_D output of the counter is connected to the C input of the 7447A. Try it and explain how it works.

For Further Investigation

Construct the circuit shown in Figure 27–1. To save you time, pin numsbers are included on the drawings. After power is applied, the read/write switch is

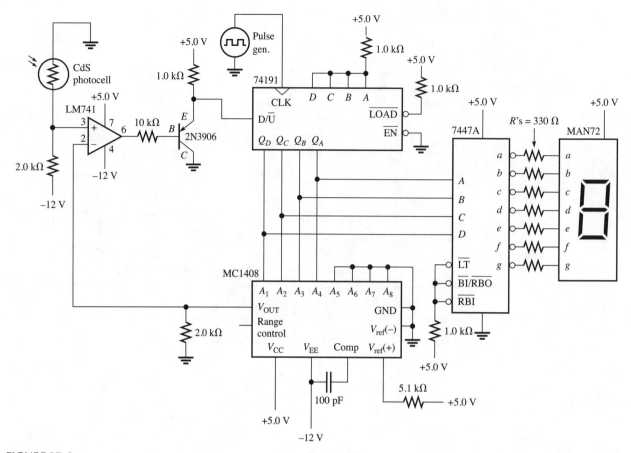

FIGURE 27–3
Simple digital light meter.

temporarily placed in the write position, then moved to the read position. As you rotate the potentiometer, the output numbers should change. Record the input voltage at each threshold. Is the output a linear function of the input voltage?

FIGURE 27–4

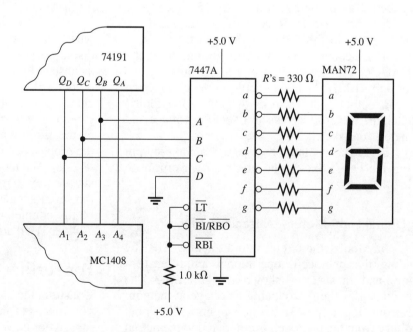

Report for Experiment 27

Name: _____ Date: _____ Class: _____

Objectives:

☐ Construct a circuit using a binary counter and a digital-to-analog converter (DAC).
☐ Check the calibration of the circuit, measure its resolution in volts/step, and observe its waveforms on an oscilloscope.
☐ Modify the circuit into a tracking analog-to-digital converter (ADC), and use it to construct a digital light meter.

Data and Observations:

Q_A
Q_B
Q_C
Q_D

PLOT 1

Q_A
Q_B
Q_C
Q_D

PLOT 2

TABLE 27–1

Quantity	Measured Value
DAC full-scale output voltage	
Volts/step	

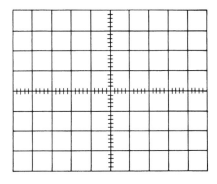

PLOT 3

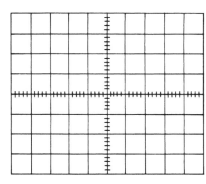

PLOT 4

207

Observations from Step 7:

Observations from Step 8:

Results and Conclusion:

Further Investigation Results:

Evaluation and Review Questions

1. a. In Step 4, you observed the analog output from the DAC. What output voltage represents a binary 1010 input?

   b. What is the output frequency of the DAC?

2. If all eight input bits for the DAC were used, what would be the output resolution in volts/step?

3. **a.** What happens to the output voltage if the 5.1 kΩ resistor is made smaller?

   b. What happens to the output resolution?

4. Explain why the circuit in Figure 27–3 exhibits oscillating behavior.

5. **a.** How could you decrease the size of the voltage steps from the circuit in Figure 27–3?

   b. Would the decrease change the sensitivity of the light meter?

6. Discuss one advantage and one disadvantage of operating the digital light meter with a higher clock frequency.

Experiment 28
Introduction to the Intel Processors

Objectives

After completing this experiment, you will be able to

☐ Use certain Debug commands to view and change internal registers, view and modify memory contents, compare a block of data, and determine the hardware configuration of the computer.

☐ Use Debug to observe selected assembly language instructions execute.

☐ Assemble and execute a simple assembly language program.

Materials Needed

PC

Summary of Theory

The Intel microprocessors are widely used as the heart of computers. The fundamental processor that is a good starting point for understanding microprocessors is the 8086/8088 processors that were introduced in 1978. As newer processors were developed, the family became known as the 80X86 and Pentium processors and later as Intel Core processors. Although significant improvements have occurred in speed, size, and structure, the concepts of how the processor performs its job are much the same. Code originally written for the 8086/8088 will still execute on newer Intel processors.

Computer programmers that use assembly language are primarily concerned with the register structure of the processor and the instruction set that is available. The register structure is the set of registers made available for the programmer to use for various functions and they often have special functions associated with them. The register structure for processors starting with the 80386 is an expanded version of the original 8086 registers. Figure 28–1 is the register structure for Intel processors.

The instruction set is the list of instructions available for programming. Most of the Intel microprocessors have complex instruction sets, with literally hundreds of variations. This lab exercise will introduce you to the registers and some basic instructions to allow you to execute an assembly language program.

A program that is available on all DOS-based computers is called Debug. Debug is a DOS program that allows you to test and debug executable files. Although Debug only allows you to work with the basic 16-bit register set, it is still a useful program for basic programming tasks such as viewing registers, testing simple programs or making changes to memory. It is a good starting point for understanding more complex assemblers like Microsoft Assembler (MASM). Table 28–1 shows the common Debug commands with examples of each. To better acquaint you with Debug and assembly language, you will use most of these commands in this lab exercise.

FIGURE 28–1
Register structure of the Intel processors.

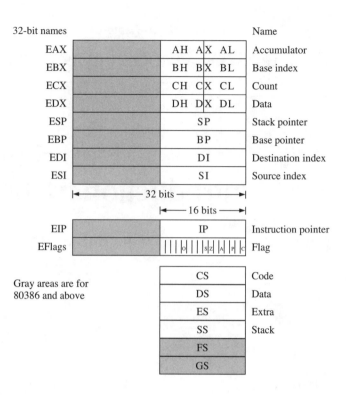

This experiment is in the form of a tutorial and does not require building hardware; it is more like a tutorial exercise than a traditional lab experiment. The procedure gives many examples of Debug commands with short exercises. Instead of a separate Report section, answers to exercises are entered in the space provided in the Procedure section.

Procedure

1. The Debug program is normally found in the DOS directory. On your PC, go into DOS and type **Debug <cr>** at the DOS prompt. (**<cr>** stands for "carriage return," which means to press the ⌷Enter⌷ key). You may need to enter the drive and DOS directory command to access Debug (as **C:\DOS>Debug <cr>.** You should see a hyphen (or minus sign) (-) which is the Debug prompt.

Debug is a versatile program that responds to a number of Debug commands as mentioned in the Summary of Theory. You can see the complete list of Debug commands if you type **?** at the Debug prompt. Try this now, and then you will be directed to use some of these commands to look at memory, the registers, and to execute some instructions. For this experiment, Debug commands that you will type will be shown in boldface; responses from the

computer are shown in normal print. All values in Debug are shown in hex, so an H is added as a suffix to the number. Since Debug expects numbers in hex format, the H is not used in entering numbers in Debug (but is often required in assemblers such as MASM).

Checking System Equipment and ROM BIOS Date

2. You can use Debug to view the contents of selected memory locations within the first 1 M of memory and pick out some of the system data for your computer. To do this, use the Dump command to first look at the BIOS data area (part of RAM) as follows:

-D 40:0 <cr>

This tells Debug to dump (display) 8 lines with 16 bytes per line of the memory starting at address 40:0 (in standard segment:offset notation). The results will be similar to the display shown in Figure 28–2(a). To the left side of the display is the address of the first byte in the line, followed by 16 bytes of hexadecimal data that are the contents of the next 16 locations. On the far right is the ASCII representation of those bytes that contain characters that can be displayed.

TABLE 28–1

Common Debug commands.

Debug Command	Examples	Description
A (assemble)	A 100 A 1234:100	Assemble instructions starting at CS:100 Assemble instructions starting at 1234:100
C (compare)	C 200,2FF,600	Compare a block of code from DS:200 to DS:2FF with the block starting at DS:600
D (dump)	D 120 D SS:FFEE FFFD	Dump (or "display") the contents of 128 bytes of memory starting at address DS:120 (DS is default) Dump 16 bytes starting at SS:FFEE
E (enter)	E 120 E:0100:0000	Examine or enter new data into memory starting at DS:120 Examine or enter new data into memory starting at 0100:0000
F (fill)	F 300, 400,FF	Fill memory locations starting at DS:300 and ending at DS:400 with the value FFH
G (go)	G = 100,120	Go (execute instructions) starting at CS:100 and continuing to CS:120. (Note equal sign is required).
H (hex)	H A000, B000	Hexadecimal sum and difference is performed on the numbers A000H and B000H
I (input)	I 300	Input port number 300H (prototype card location)
M (move)	M 200,2FF,600	Move (copy) the block of code from DS:200 to DS:2FF to the block starting at DS:600
O (output)	O 43,36	Out the number 36H to port 43 (timer control register)
P (proceed)	P = 100 10	Proceed, or execute a set of related instructions starting at address CS:100 and doing 10 instructions. It can be used for loops or calls to subroutines as if it were a single instruction.
Q (quit)	Q	Quit Debug
R (register)	R R SI	Display the contents of the 16-bit registers and flags Display the contents of the SI register and allow a new value to be written to it
S (search)	S 100, FFF, 0A	Search locations DS:100 to DS:FFF for the byte 0AH. Address with this byte will be displayed.
T (trace)	T T 5	Trace the execution of one instruction Trace the execution of the next 5 instructions
U (unassemble)	U 100 U 100 110	Unassemble (disassemble) code starting at location CS:100 Unassemble code from CS:100 to CS:110

The data that is shown as a result of this dump command is a "work area" for BIOS and contains lots of information about your computer. For example, location 40:17 has keyboard control information. Bit 7 (the most significant bit) is a 1 if the insert mode is on and bit 6 will indicate if the *caps lock* function is on. Press the caps lock button and repeat the dump command as before. You should observe bit 6 of 40:17 will now be set as shown in Figure 28–2(b). (Reset the caps lock function after observing this.)

Now observe the first two locations at 40:10. These locations contain the equipment status word. To interpret the information, you have to reverse the two bytes and convert them to binary. For example, in Figure 28–2(b) the equipment status word is 27 C2 (read on the second line). Intel shows words in reverse order; the most significant byte is first

```
         C:\WINDOWS>Debug
         -D 40:0
         0040:0000  F8 03 F8 02 00 00 00 00-78 03 00 00 00 00 0F 02    ........x.......
         0040:0010  27 C2 00 80 02 00 00 20-00 00 38 00 38 00 44 20    '...... ..8.8.D
         0040:0020  65 12 62 30 75 16 67 22-0D 1C 44 20 20 39 34 05    e.b0u.g"..D  94.
         0040:0030  30 0B 3A 27 30 0B 0D 1C-00 00 00 00 00 00 00 C0    0.:'0...........
         0040:0040  00 01 80 00 00 00 00 00-00 03 50 00 00 10 00 00    ..........P.....
         0040:0050  00 0C 00 00 00 00 00 00-00 00 00 00 00 00 00 00    ................
         0040:0060  0E 0D 00 D4 03 29 30 A4-17 3D 85 00 CE 61 0A 00    .....)0..=...a..
         0040:0070  00 00 00 00 00 01 00 00-14 14 14 3C 01 01 01 01    ...........<....
```

(a)

```
         -D 40:0
         0040:0000  F8 03 F8 02 00 00 00 00-78 03 00 00 00 00 0F 02    ........x.......
         0040:0010  27 C2 00 80 02 00 00 60-00 00 2C 00 2C 00 44 20    '......`..,.,.D
         0040:0020  20 39 34 05 30 0B 3A 27-30 0B 0D 1C 20 39 34 05     94.0.:'0... 94.
         0040:0030  30 0B 3A 27 30 0B 0D 1C-0D 1C 0D 1C 0D 1C 00 C0    0.:'0...........
         0040:0040  00 01 80 00 00 00 00 00-00 03 50 00 00 10 00 00    ..........P.....
         0040:0050  00 18 00 00 00 00 00 00-00 00 00 00 00 00 00 00    ................
         0040:0060  0E 0D 00 D4 03 29 30 A4-17 3D 85 00 72 63 0A 00    .....)0..=..rc..
         0040:0070  00 00 00 00 00 01 00 00-14 14 14 3C 01 01 01 01    ...........<....
```

(b)

FIGURE 28–2

followed by the least significant byte. Reversing the bytes and converting to binary results in the following:

Bit:	15	14	13	12	11	10	9	8	7	6	5	4	3	2	1	0
Binary:	1	1	0	0	0	0	1	0	0	0	1	0	0	1	1	1

From left to right, this pattern means

15, 14   Number of parallel printer ports attached = 3 (binary 11)

13, 12   Not used

11–9   Number of serial ports attached = 1 (binary 001)

8   Not used

7, 6   Number of diskette devices = 1 (where 00 = 1, 01 = 2, 10 = 3, and 11 = 4)

5, 4   Initial video mode = 80×25 color (where 01 = 40×25 color, 10 = 80×25 color, and 11 = 80×25 monochrome)

3, 2   Not used

1   Math coprocessor is present = yes (0 = no, 1 = yes)

0   Diskette drive is present = yes (0 = no, 1 = yes)

In the space provided, write down the bit pattern you see for locations 40:10 and 40:11 in reverse order. Then determine what installed equipment BIOS has found.

Bit:	15	14	13	12	11	10	9	8	7	6	5	4	3	2	1	0
Binary:																

Number of parallel printer ports attached =

_____

Number of serial ports attached =

_____

Number of diskette devices =

_____

Initial video mode = _____

Math coprocessor is present? _____

Diskette drive is present? _____

3. In this step, you can check the date of manufacture of your ROM BIOS recorded as mm/dd/yy which begins at location FFFF:5. Issue a dump command as follows:

-D FFFF: 0 <cr>

The date is encoded as ASCII characters which can be read directly from the right-hand side of the display (starting at offset 5). Indicate the date you found in the space provided:

Date of ROM BIOS:_____

Changing Data in Memory and Comparing a Block

4. In this step, you will change the contents of 256 memory locations to an ASCII zero (30H) by using the Fill command. First check on the contents of the current data segment between offset address 0 and 100 by using the Dump command as follows:

-D 0 100 <cr>

You should observe a random pattern of bytes in the memory. Now fill this region with ASCII zero (30H) by invoking the Fill command as follows:

-F 0,FF,30 <cr>

Next issue the Dump command again and observe the results. Notice the pattern on the right side of the display.

5. Now change one location to a different character by invoking the Enter command as follows:

-E 20 <cr>

Debug will respond with the address (DS:20) and the contents (30) followed by a dot. Enter **31 <cr>** after the dot. This will change the byte at location DS:20 to an ASCII 1.
(You can confirm this with another Dump command.)

6. Issue the following command which will compare two blocks of memory:

-C 0,2F,30 <cr>

Describe the response from Debug.

_____

Examining and Modifying the Internal Registers

7. Issue the register command by typing:

-R <cr>

You should see the contents of the 16 bit register set with their initial values similar to Figure 28–3. The general purpose registers (AX, BX, CX, and DX), the base pointer (BP), and the source and destination index registers (SI and DI) will all show 0000H. The segment registers (DS, ES, SS, and CS) will have a hex number that depends on where DOS found available address space so will not be the same for different users. The stack pointer (SP) begins near the top of the offset address (FFEE) and the instruction pointer will start at 0100H. Flags are indicated with two-letter codes. Following the register list is an address given in segment:offset notation and an arbitrary instruction.

8. To modify the contents of the AX register, issue the register command for the AX register:

-R AX <cr>

The processor will respond with

AX 0000:

showing the current contents and a colon. The colon indicates it is waiting for your response. Type

:0100 <cr>

The value 0100H will now be entered into the AX register. You can see this by issuing another register command. Type

-R <cr>

and you should see the new contents of the AX register. The other register can be modified by a similar procedure. Use the register command to change the contents of the BX register to 0200H and the CX register to F003H. Indicate in the space below the command you used:

Command to place 0200H in BX register:

_____

Command to place F003H in CX register:

_____

Tracing Instructions

9. In this step, you will assemble some basic assembly language instructions with Debug and

```
-R
AX=0000  BX=0000  CX=0000  DX=0000  SP=FFEE  BP=0000  SI=0000  DI=0000
DS=1E8E  ES=1E8E  SS=1E8E  CS=1E8E  IP=0100     NV UP EI PL NZ NA PO NC
1E8E:0100 BAAB81        MOV     DX,81AB
-
```

FIGURE 28–3

then observe the results after they execute. Issue the following Debug command:

-A 100 <cr>

This command tells Debug to assemble instructions beginning at CS:100. The processor responds with the starting address and a blinking dash cursor. Enter the following assembly language instructions. Note that the segment address will be different than the one shown here, but the offset address should be the same.

> 1E8C:0100 **mov cx, 04 <cr>**
> 1E8C:0103 **mov ax, 40 <cr>**
> 1E8C:0106 **mov ds,ax <cr>**
> 1E8C:0108 **mov si, 17 <cr>**
> 1E8C:010B **xor byte ptr [si], 40 <cr>**
> 1E8C:010E **loop 10B <cr>**
> 1E8C:0110 **nop <cr>**

Press **<cr>** again to leave the assembler.

10. You can trace this code and observe the registers as each instruction is executed. The first four instructions are all move instructions that are used to preset certain values in registers. An explanation of the code follows. First, make sure the IP is loaded with 0100H. If not, use a **R IP** instruction to set it to the starting address of the code. Then use the -**T** command to trace a single instruction. You should observe that after executing this instruction, the CX register contains 0004H.

Issue three more -**T** commands until you reach the **xor byte ptr [si], 40** instruction. Up to this point you should observe that certain registers are loaded. The register pair DS:SI forms a segment:offset address which points to the location in memory you looked at in step 2. This memory location is the BIOS work area with the keyboard control information. Recall that bit 6 will indicate if the *caps lock* function is on. You may also recall that an XOR gate can be used to pass a bit unchanged or to complement it.* In the instruction, the XOR function is used to selectively complement bit 6 and leave the others unchanged at the location pointed to by DS:SI. Issue another trace command, observe what happens after the XOR instruction executes, and record your observation in the space provided:

_____

_____

11. Following the XOR instruction is the **loop 10B** instruction. The loop instruction is a quick way to repeat a process. Loop decrements the CX register by 1 and transfers control to the instruction at

216

location 10B if CX is not 0; otherwise, the next instruction in line is executed. Since the CX register was initialized to 4, the loop will be repeated 4 times. Continue tracing instructions, watching what happens, until you reach the **nop** ("no operation") instruction on line CS:110. *Note:* if you want to repeat the procedure outlined in this step, you need to reset the IP to 100. You can then trace the program again. Observations from step 11:

_____

_____

Assembling and Executing an Assembly Language Program

12. The program for this step is very similar to Example 13–2 given in the text. To make it interesting, we'll change how the data is entered and make it byte-size (instead of word-size). This will require changing several instructions in the code. The problem is to write an assembly language program that will find the largest unsigned number

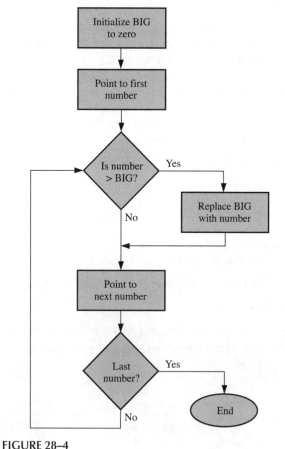

FIGURE 28–4
Flowchart for largest number problem. The variable BIG represents the largest value.

in a list and place it in the last position. Data in this exercise is bytes (8 bits), not words (16 bits). The last data point is signaled with a zero. The flowchart is shown in Figure 28–4 for reference.*

You can use Debug to enter the data into the list. The list will be 16 data points starting at the current data segment (assigned by DOS) and at the offset address of 50. Select 16 random bytes (not in any order) and modify memory by using the Debug Enter command starting at DS:50 (-**E 50**). You can use the space bar after each entry until you reach DS:60. Add zero as a last data indicator for the 17th byte at address DS:60, then press **<cr>**. Check that the data has been correctly entered with the Debug Dump command.

13. The Debug listing of the modified program is given in this step. The program is entered by starting the assembly at location 100. Start by issuing the -**A 100** command. Then type the following. (Note that your segment address will be different).

```
1E8C:0100 mov ax,0 <cr>
1E8C:0103 mov bx,50 <cr>
1E8C:0106 cmp [bx],al <cr>
1E8C:0108 jbe 010c <cr>
1E8C:010A mov al,[bx] <cr>
1E8C:010C inc bx <cr>
1E8C:010D cmp byte ptr [bx],0 <cr>
1E8C:0110 jnz 0106 <cr>
1E8C:0112 mov [bx],al <cr>
1E8C:0114 nop <cr>
```

---

*Although the flowchart is identical to the one in Figure 13-33 of the text, the instructions are not because the data are bytes, not words.

Check that you have entered the code correctly by unassembling it using the **u** command:

-u 100 114

If the code has been correctly entered, you can execute it with the Go command. (Alternatively, you can trace each instruction if you prefer). Issue the Go command as follows. (Note the = sign!)

-g = 100 114 <cr>

The code will execute and show the registers at the end of the code. Indicate in the space below the values in the three registers listed after executing the code:

AX = _____ BX = _____

CX = _____

Press **<cr>** again to leave the assembler.

14. Dump the memory locations from DS:50 to DS:60. What happened?

_____

_____

For Further Investigation

For the code in step 13, try changing the instruction on line 100 from **mov ax,0** to **mov ax,ffff** and the instruction on line 108 from **jbe 010c** to **jae 010c.** Reenter a zero for the last data point (at DS:50). Run the code and observe the data and the registers. Summarize your findings.

Evaluation and Review Questions

1. Cite evidence from the experiment that the Debug Dump command can be used for looking at both RAM and ROM.

2. Explain what action occurs as a result of the Debug command -**F 1F0,1FF,FA <cr>.**

3. Show the Debug command that would allow you to view the lowest 10 words in memory containing the Intel interrupt vectors.

4. Where is the BIOS "work area" in memory? Cite examples of the type of data that can be found there.

5. The assembly language instruction XOR AX,AX causes the AX register to be cleared. Explain why.

6. Explain what happens with the assembly language LOOP instruction.

Experiment 29
Application of Bus Systems

Objectives

After completing this experiment, you will be able to

☐ Build a liquid-level detector and display that are interconnected with a bus system.

☐ Connect your system with that of another student. Assign each detector a unique address.

☐ Design and construct a decoder circuit that selects the desired detector and sends its data to the displays.

Materials Needed

14051B analog MUX/DMUX
14532B priority encoder
555 timer
Resistors: one 330 Ω, six 1.0 kΩ, six 27 kΩ, one 200 kΩ, eight 4.7 MΩ
Capacitors: one 0.01 µF, two 0.1 µF, one 1.0 µF
Ten DIP switches (wire jumpers may be substituted)
Four 2N3904 NPN transistors (or equivalent)
Eight LEDs
Other materials as determined by student

Summary of Theory

In many digital systems, various parts of the system must be able to exchange data with one another. These devices include computers and pe-

ripherals, memories, and electronic instruments such as voltmeters and signal generators. To interconnect each of these devices with its own set of wires for exchange of data would result in unnecessary complexity, to say the least. Instead, they are connected to a common set of wires known as a *data bus*.

In a bus system, only one device can send data at one time, although more than one device may receive data. A *protocol* (who can "talk" and who should "listen") needs to be established when the system is designed. Signals used to control the movement of data are sent over a separate bus called a *control bus*.

Another type of bus used with computers is called an *address bus*. This bus carries address information that selects a particular location in memory or a device.

To avoid bus conflicts, logic devices that drive buses must be able to be disconnected from the bus when they are not "talking." There are two methods for achieving this. The first uses *open-collector logic* in which the upper output transistor is simply omitted from the totem-pole (see Figure 29–1). A single external pull-up resistor is attached to each bus line, which serves as the load for all open-collector devices connected to that line. For example, if several open-collector NAND gates are connected to the line, any one of them can pull the line LOW. Active command signals are usually LOW so that various devices can be added to or removed from the bus without affecting it.

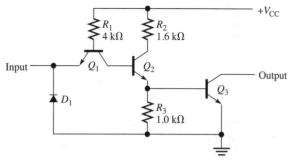

(a) Open-collector inverter circuit

FIGURE 29–1

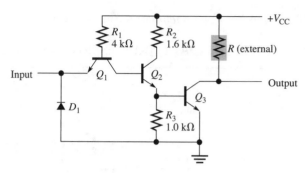

(b) With external pull-up resistor

The second method, which is faster and less subject to noise, uses *three-state logic* (also known as Tristate logic, a trademark of National Semiconductor Corporation). Tristate logic uses a separate enable line to connect the device to the bus. When it is not enabled, it is in a high-impedance or floating state. Pull-up resistors are not needed as they are in open-collector logic.

Actually, open-collector logic can be implemented with ICs that are not special open-collector types by using discrete transistors for the open-collector device. The advantage of this technique is the added drive current that can be supplied (but with additional complexity and cost). This method is illustrated in this experiment.

The circuit shown in Figure 29–2 represents a liquid-level-sensing circuit for a remote "container." The liquid-level data are sensed by a series of switches that close, one after the other, as the liquid level rises. The data are transmitted over a data bus to a display circuit, which causes an LED to indicate the level. The transistors are connected as open-collector devices using R_{17}, R_{18}, R_{19}, and R_{20} as the collector resistors. Only one collector resistor is necessary for each line on the data bus, even if more "containers" are added. The discrete transistors act as electronic switches. A logic LOW from the priority encoder keeps the transistors off (an open switch) and causes the bus line to be pulled HIGH by the pull-up resistor; a HIGH from the priority encoder turns the transistors on (closed switch) causing a LOW to appear on the bus line. The transistors thus invert the logic that goes to the data bus. The transistors allow added drive current for the data lines and convert the priority encoder to an open-collector device, thus allowing the addition of other sensors on the same bus. The address bus is shown with no connections other than switches; you

will design a circuit to connect the address bus to the priority encoder in this experiment.

The 14051B analog MUX/DMUX chosen for the display is an interesting and versatile IC. As you can see from the data sheet, this IC functions as a set of digitally controlled switches. The common out/in line is connected on one side to all switches. One switch is closed by the control line; the remaining switches are left open. This process allows an analog signal to be routed from either the common side or the switch side. The circuit in Figure 29–2 makes use of this capability by sending a clock signal from the 555 timer to blink the light, which indicates the correct level.

Procedure

1. Connect the liquid-level-sensing circuit shown in Figure 29–2.* The bus wires are shown with suggested wire colors to make interfacing with another student's experiment easier. Note that the address lines are not connected in this step.

2. Test your circuit by closing switches S_0 through S_7 to simulate a rising liquid level. The LED indicators should follow the switches. If no switch is closed, all LEDs are off.

3. Test the effect of the GS line on the circuit. Temporarily remove the connection from the GS bus line to the inhibit (pin 6) of the 14051B and tie the inhibit line to a LOW with all switches turned off. The explanation of the effect you see can be found by observing closely the second line on the 14532B truth table and the next-to-last line of the

---

*If you need to simplify the circuit, the 555 timer can be left out. Connect +5.0 V to the left side of R_{16}.

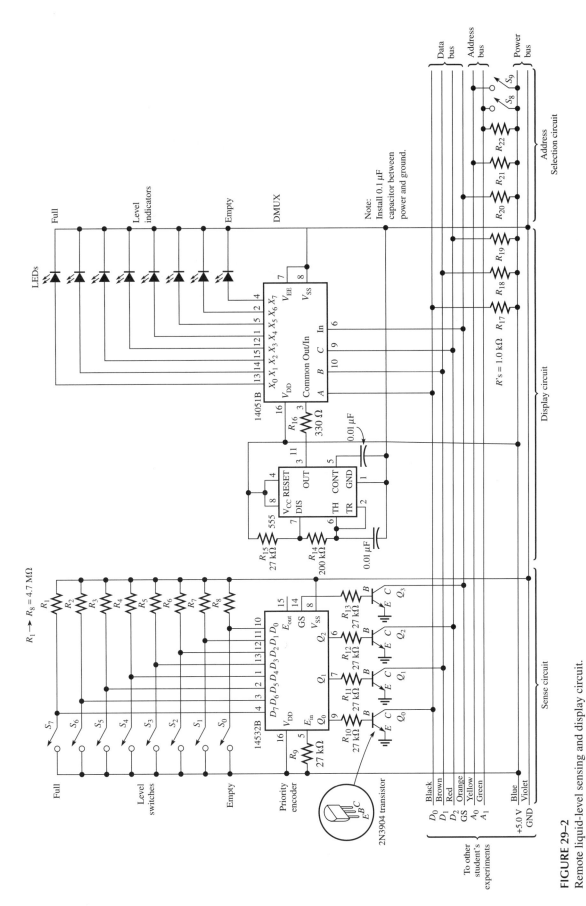

FIGURE 29–2

Remote liquid-level sensing and display circuit.

221

TABLE 29–1
Truth table for MC14051B.

Control Inputs				ON Switches					
	Select								
Inhibit	C*	B	A	MC14051B	MC14052B		MC14053B		
0	0	0	0	X0	Y0	X0	Z0	Y0	X0
0	0	0	1	X1	Y1	X1	Z0	Y0	X1
0	0	1	0	X2	Y2	X2	Z0	Y1	X0
0	0	1	1	X3	Y3	X3	Z0	Y1	X1
0	1	0	0	X4			Z1	Y0	X0
0	1	0	1	X5			Z1	Y0	X1
0	1	1	0	X6			Z1	Y1	X0
0	1	1	1	X7			Z1	Y1	X1
1	X	X	X	None	None		None		

*Not applicable for MC14052
X = Don't Care

TABLE 29–2
Truth table for MC14532B.

Input									Output				
E_{in}	D7	D6	D5	D4	D3	D2	D1	D0	GS	Q2	Q1	Q0	E_{out}
0	X	X	X	X	X	X	X	X	0	0	0	0	0
1	0	0	0	0	0	0	0	0	0	0	0	0	1
1	1	X	X	X	X	X	X	X	1	1	1	1	0
1	0	1	X	X	X	X	X	X	1	1	1	0	0
1	0	0	1	X	X	X	X	X	1	1	0	1	0
1	0	0	0	1	X	X	X	X	1	1	0	0	0
1	0	0	0	0	1	X	X	X	1	0	1	1	0
1	0	0	0	0	0	1	X	X	1	0	1	0	0
1	0	0	0	0	0	0	1	X	1	0	0	1	0
1	0	0	0	0	0	0	0	1	1	0	0	0	0

X = Don't Care

14051B truth table. The truth tables are given as Table 29–1 and Table 29–2. (Remember that the transistor inverts the logic.) Then restore the circuit to its original condition. Summarize your observations of this step in the report.

4. In this step you will connect your bus lines with those of another student or two. Only one student should keep the address selection circuit and the pull-up resistors R_{17} through R_{22}. Other students connected to the same bus should remove these resistors and the address selection switches. Also, power and ground must be supplied from only one power supply. Select a unique address for each student's encoder on the bus and design a circuit to decode each address.* Use the E_{in} (enable input) on the 14532B as a "chip select" to place the selected encoder on the bus. Draw the schematic for your decoder and the other decoders using the common bus in the report.

5. If your circuit is working properly, you should be able to set an address on the address switches (S_8 and S_9) and be able to see the "liquid" level (switch settings) for the selected container (encoder). All displays should show the same container. Demonstrate your working circuit to your instructor.

---

*If you use a TTL circuit to interface to the 14532, connect a 3.3 kΩ resistor between the TTL output and +5.0 V to help pull up the voltage; do not connect any other load to the TTL output. A 7400 can be used for each decoder.

For Further Investigation

The circuit in this experiment can be cycled through the addresses you have selected by adding a counter to the address line. Design a counter that selects the four possible addresses for the address bus. The counter should indicate which address is being selected at any time in a seven-segment display. If a selected address is not used, all the level indicator LEDs should be out. Draw your schematic in your report.

Alternate Further Investigation

If you have computer programming experience, you can connect the address bus to an output port of a computer and program the computer to read each liquid level on an input port.

Report for Experiment 29

Name: _____ Date: _____ Class: _____

Objectives:

☐ Build a liquid-level detector and display that are interconnected with a bus system.
☐ Connect your system with that of another student. Assign each detector a unique address.
☐ Design and construct a decoder circuit that selects the desired detector and sends its data to the displays.

Data and Observations:

Observations from Step 3:

Decoder schematic:

Results and Conclusion:

Further Investigation Results:

Evaluation and Review Questions

1. a. What is the purpose of the transistors for the circuit in Figure 29–2?

    b. Why is it unnecessary to have a tristate or open-collector device on the display circuit?

2. What assures that there is no conflict on the data bus with several different priority encoders connected to the same lines?

3. The circuit in this experiment shows the 14051B connected as a DMUX. Assume you wanted to use the same IC in a circuit where you needed a MUX. Explain where you would connect the inputs and the output.

4. What is the purpose of R_{16}? What happens if it were to open?

5. Assume the output from the 555 timer in Figure 29–2 is a constant HIGH.
    a. What effect would this have on the LED level indicators?

    b. What is the effect on the LED level indicators if the 555 output is shorted to ground?

6. Assume that two level indicators on the same bus try to send their data to the indicator at the same time. (This is called a *bus contention error*). One of the indicators is trying to send a logic LOW; the other is trying to send a logic HIGH.
    a. Who wins?

    b. Why?

    c. How would you troubleshoot the circuit to determine which level indicator *should* be on the bus?

Appendix A

Manufacturers' Data Sheets

DM5400/DM7400 Quad 2-Input NAND Gates

General Description

This device contains four independent gates each of which performs the logic NAND function.

Absolute Maximum Ratings (Note 1)

Supply Voltage	7V
Input Voltage	5.5V
Storage Temperature Range	−65°C to 150°C

Note 1: The "Absolute Maximum Ratings" are those values beyond which the safety of the device can not be guaranteed. The device should not be operated at these limits. The parametric values defined in the "Electrical Characteristics" table are not guaranteed at the absolute maximum ratings. The "Recommended Operating Conditions" table will define the conditions for actual device operation.

Connection Diagram

Dual-In-Line Package

DM5400 (J)   DM7400 (N)

TL/F/6613-1

Function Table

$Y = \overline{AB}$

Inputs		Output
A	B	Y
L	L	H
L	H	H
H	L	H
H	H	L

H = High Logic Level
L = Low Logic Level

Recommended Operating Conditions

Symbol	Parameter	DM5400			DM7400			Units
		Min	Nom	Max	Min	Nom	Max	
V_{CC}	Supply Voltage	4.5	5	5.5	4.75	5	5.25	V
V_{IH}	High Level Input Voltage	2			2			V
V_{IL}	Low Level Input Voltage			0.8			0.8	V
I_{OH}	High Level Output Current			−0.4			−0.4	mA
I_{OL}	Low Level Output Current			16			16	mA
T_A	Free Air Operating Temperature	−55		125	0		70	°C

Electrical Characteristics over recommended operating free air temperature (unless otherwise noted)

Symbol	Parameter	Conditions		Min	Typ (Note 1)	Max	Units
V_I	Input Clamp Voltage	$V_{CC} = $ Min, $I_I = -12$ mA				−1.5	V
V_{OH}	High Level Output Voltage	$V_{CC} = $ Min, $I_{OH} = $ Max $V_{IL} = $ Max		2.4	3.4		V
V_{OL}	Low Level Output Voltage	$V_{CC} = $ Min, $I_{OL} = $ Max $V_{IH} = $ Min			0.2	0.4	V
I_I	Input Current @ Max Input Voltage	$V_{CC} = $ Max, $V_I = 5.5$V				1	mA
I_{IH}	High Level Input Current	$V_{CC} = $ Max, $V_I = 2.4$V				40	µA
I_{IL}	Low Level Input Current	$V_{CC} = $ Max, $V_I = 0.4$V				−1.6	mA
I_{OS}	Short Circuit Output Current	$V_{CC} = $ Max (Note 2)	DM54	−20		−55	mA
			DM74	−18		−55	
I_{CCH}	Supply Current With Outputs High	$V_{CC} = $ Max			4	8	mA
I_{CCL}	Supply Current With Outputs Low	$V_{CC} = $ Max			12	22	mA

Switching Characteristics at $V_{CC} = 5$V and $T_A = 25$°C (See Section 1 for Test Waveforms and Output Load)

Parameter	Conditions	$C_L = 15$ pF $R_L = 400\Omega$			Units
		Min	Typ	Max	
t_{PLH} Propagation Delay Time Low to High Level Output			12	22	ns
t_{PHL} Propagation Delay Time High to Low Level Output			7	15	ns

Note 1: All typicals are at $V_{CC} = 5$V, $T_A = 25$°C.
Note 2: Not more than one output should be shorted at a time.

National Semiconductor

DM5402/DM7402 Quad 2-Input NOR Gates

General Description

This device contains four independent gates each of which performs the logic NOR function.

Absolute Maximum Ratings (Note 1)

Supply Voltage	7V
Input Voltage	5.5V
Storage Temperature Range	−65°C to 150°C

Note 1: The "Absolute Maximum Ratings" are those values beyond which the safety of the device can not be guaranteed. The device should not be operated at these limits. The parametric values defined in the "Electrical Characteristics" table are not guaranteed at the absolute maximum ratings. The "Recommended Operating Conditions" table will define the conditions for actual device operation.

Connection Diagram

Dual-In-Line Package

TL/F/6492-1

DM5402 (J)   DM7402 (N)

Function Table

$$Y = \overline{A + B}$$

Inputs		Output
A	B	Y
L	L	H
L	H	L
H	L	L
H	H	L

H = High Logic Level
L = Low Logic Level

Absolute Maximum Ratings (Note)

If Military/Aerospace specified devices are required, please contact the National Semiconductor Sales Office/Distributors for availability and specifications.

Supply Voltage	7V
Input Voltage	5.5V
Operating Free Air Temperature Range	
DM54 and 54	−55°C to +125°C
DM74	0°C to +70°C
Storage Temperature Range	−65°C to +150°C

Note: The "Absolute Maximum Ratings" are those values beyond which the safety of the device cannot be guaranteed. The device should not be operated at these limits. The parametric values defined in the "Electrical Characteristics" table are not guaranteed at the absolute maximum ratings. The "Recommended Operating Conditions" table will define the conditions for actual device operation.

Recommended Operating Conditions

Symbol	Parameter	DM5402 Min	DM5402 Nom	DM5402 Max	DM7402 Min	DM7402 Nom	DM7402 Max	Units
V_{CC}	Supply Voltage	4.5	5	5.5	4.75	5	5.25	V
V_{IH}	High Level Input Voltage	2			2			V
V_{IL}	Low Level Input Voltage			0.8			0.8	V
I_{OH}	High Level Output Current			−0.4			−0.4	mA
I_{OL}	Low Level Output Current			16			16	mA
T_A	Free Air Operating Temperature	−55		125	0		70	°C

Electrical Characteristics
over recommended operating free air temperature range (unless otherwise noted)

Symbol	Parameter	Conditions	Min	Typ (Note 1)	Max	Units
V_I	Input Clamp Voltage	$V_{CC} = Min$, $I_I = -12$ mA			−1.5	V
V_{OH}	High Level Output Voltage	$V_{CC} = Min$, $I_{OH} = Max$ $V_{IL} = Max$	2.4	3.4		V
V_{OL}	Low Level Output Voltage	$V_{CC} = Min$, $I_{OL} = Max$ $V_{IH} = Min$		0.2	0.4	V
I_I	Input Current @ Max Input Voltage	$V_{CC} = Max$, $V_I = 5.5V$			1	mA
I_{IH}	High Level Input Current	$V_{CC} = Max$, $V_I = 2.4V$			40	µA
I_{IL}	Low Level Input Current	$V_{CC} = Max$, $V_I = 0.4V$			−1.6	mA
I_{OS}	Short Circuit Output Current	$V_{CC} = Max$ (Note 2) DM54	−20		−55	mA
		DM74	−18		−55	mA
I_{CCH}	Supply Current with Outputs High	$V_{CC} = Max$		8	16	mA
I_{CCL}	Supply Current with Outputs Low	$V_{CC} = Max$		14	27	mA

Switching Characteristics at $V_{CC} = 5V$ and $T_A = 25°C$ (See Section 1 for Test Waveforms and Output Load)

Symbol	Parameter	Conditions	Min	Max	Units
t_{PLH}	Propagation Delay Time Low to High Level Output	$C_L = 15$ pF $R_L = 400\Omega$		22	ns
t_{PHL}	Propagation Delay Time High to Low Level Output			15	ns

Note 1: All typicals are at $V_{CC} = 5V$, $T_A = 25°C$.
Note 2: Not more than one output should be shorted at a time.

National Semiconductor

DM5404/DM7404 Hex Inverting Gates

General Description

This device contains six independent gates each of which performs the logic INVERT function.

Absolute Maximum Ratings (Note 1)

Supply Voltage	7V
Input Voltage	5.5V
Storage Temperature Range	−65°C to 150°C

Note 1: The "Absolute Maximum Ratings" are those values beyond which the safety of the device can not be guaranteed. The device should not be operated at these limits. The parametric values defined in the "Electrical Characteristics" table are not guaranteed at the absolute maximum ratings. The "Recommended Operating Conditions" table will define the conditions for actual device operation.

Function Table

$Y = \bar{A}$

Input	Output
A	Y
L	H
H	L

H = High Logic Level
L = Low Logic Level

Connection Diagram

Dual-In-Line Package

TL/F/6494-1

DM5404 (J) DM7404 (N)

Absolute Maximum Ratings (Note)

If Military/Aerospace specified devices are required, please contact the National Semiconductor Sales Office/Distributors for availability and specifications.

Supply Voltage	7V
Input Voltage	5.5V
Operating Free Air Temperature Range	
DM54 and 54	−55°C to +125°C
DM74	0°C to +70°C
Storage Temperature Range	−65°C to +150°C

Note: The "Absolute Maximum Ratings" are those values beyond which the safety of the device cannot be guaranteed. The device should not be operated at these limits. The parametric values defined in the "Electrical Characteristics" table are not guaranteed at the absolute maximum ratings. The "Recommended Operating Conditions" table will define the conditions for actual device operation.

Recommended Operating Conditions

Symbol	Parameter	DM5402 Min	DM5402 Nom	DM5402 Max	DM7402 Min	DM7402 Nom	DM7402 Max	Units
V_{CC}	Supply Voltage	4.5	5	5.5	4.75	5	5.25	V
V_{IH}	High Level Input Voltage	2			2			V
V_{IL}	Low Level Input Voltage			0.8			0.8	V
I_{OH}	High Level Output Current			−0.4			−0.4	mA
I_{OL}	Low Level Output Current			16			16	mA
T_A	Free Air Operating Temperature	−55		125	0		70	°C

Electrical Characteristics

over recommended operating free air temperature range (unless otherwise noted)

Symbol	Parameter	Conditions	Min	Typ (Note 1)	Max	Units
V_I	Input Clamp Voltage	V_{CC} = Min, I_I = −12 mA			−1.5	V
V_{OH}	High Level Output Voltage	V_{CC} = Min, I_{OH} = Max V_{IL} = Max	2.4	3.4		V
V_{OL}	Low Level Output Voltage	V_{CC} = Min, I_{OL} = Max V_{IH} = Min		0.2	0.4	V
I_I	Input Current @ Max Input Voltage	V_{CC} = Max, V_I = 5.5V			1	mA
I_{IH}	High Level Input Current	V_{CC} = Max, V_I = 2.4V			40	µA
I_{IL}	Low Level Input Current	V_{CC} = Max, V_I = 0.4V			−1.6	mA
I_{OS}	Short Circuit Output Current	V_{CC} = Max (Note 2) DM54 / DM74	−20 / −18		−55 / −55	mA
I_{CCH}	Supply Current with Outputs High	V_{CC} = Max		8	16	mA
I_{CCL}	Supply Current with Outputs Low	V_{CC} = Max		14	27	mA

Switching Characteristics at V_{CC} = 5V and T_A = 25°C (See Section 1 for Test Waveforms and Output Load)

Symbol	Parameter	Conditions	Min	Max	Units
t_{PLH}	Propagation Delay Time Low to High Level Output	C_L = 15 pF R_L = 400Ω		22	ns
t_{PHL}	Propagation Delay Time High to Low Level Output			15	ns

Note 1: All typicals are at V_{CC} = 5V, T_A = 25°C.
Note 2: Not more than one output should be shorted at a time.

TYPES SN5408, SN7408
QUADRUPLE 2-INPUT POSITIVE-AND GATES

recommended operating conditions

		SN5408			SN7408		UNIT
	MIN	NOM	MAX	MIN	NOM	MAX	
V_{CC} Supply voltage	4.5	5	5.5	4.75	5	5.25	V
V_{IH} High-level input voltage	2			2			V
V_{IL} Low-level input voltage			0.8			0.8	V
I_{OH} High-level output current			-0.8			-0.8	mA
I_{OL} Low-level output current			16			16	mA
T_A Operating free-air temperature	-55		125	0		70	°C

electrical characteristics over recommended operating free-air temperature range (unless otherwise noted)

PARAMETER	TEST CONDITIONS†	SN5408 MIN	TYP‡	MAX	SN7408 MIN	TYP‡	MAX	UNIT
V_{IK}	V_{CC} = MIN, I_I = -12 mA			-1.5			-1.5	V
V_{OH}	V_{CC} = MIN, V_{IH} = 2 V, I_{OH} = -0.8 mA	2.4	3.4		2.4	3.4		V
V_{OL}	V_{CC} = MIN, V_{IL} = 0.8 V, I_{OL} = 16 mA		0.2	0.4		0.2	0.4	V
I_I	V_{CC} = MAX, V_I = 5.5 V			1			1	mA
I_{IH}	V_{CC} = MAX, V_I = 2.4 V			40			40	µA
I_{IL}	V_{CC} = MAX, V_I = 0.4 V			-1.6			-1.6	mA
I_{OS}§	V_{CC} = MAX	-20		-55	-18		-55	mA
I_{CCH}	V_{CC} = MAX, V_I = 4.5 V		11	21		11	21	mA
I_{CCL}	V_{CC} = MAX, V_I = 0 V		20	33		20	33	mA

† For conditions shown as MIN or MAX, use the appropriate value specified under recommended operating conditions.
‡ All typical values are at V_{CC} = 5 V, T_A = 25°C.
§ Not more than one output should be shorted at a time.

switching characteristics, V_{CC} = 5 V, T_A = 25°C (see note 2)

PARAMETER	FROM (INPUT)	TO (OUTPUT)	TEST CONDITIONS	SN7408 MIN	TYP	MAX	UNIT
t_{PLH}	A or B	Y	R_L = 400 Ω, C_L = 15 pF		17.5	27	ns
t_{PHL}					12	19	ns

NOTE 2: See General Information Section for load circuits and voltage waveforms.

TYPES SN5408, SN54LS08, SN54S08, SN7408, SN74LS08, SN74S08
QUADRUPLE 2-INPUT POSITIVE-AND GATES
REVISED DECEMBER 1983

- Package Options Include Both Plastic and Ceramic Chip Carriers in Addition to Plastic and Ceramic DIPs
- Dependable Texas Instruments Quality and Reliability

description

These devices contain four independent 2-input AND gates.

The SN5408, SN54LS08, and SN54S08 are characterized for operation over the full military temperature range of -55°C to 125°C. The SN7408, SN74LS08 and SN74S08 are characterized for operation from 0°C to 70°C.

SN5408, SN54LS08, SN54S08 ... J OR W PACKAGE
SN7408 ... J OR N PACKAGE
SN74LS08, SN74S08 ... D, J OR N PACKAGE
(TOP VIEW)

```
1A  [1   14] VCC
1B  [2   13] 4B
1Y  [3   12] 4A
2A  [4   11] 4Y
2B  [5   10] 3B
2Y  [6    9] 3A
GND [7    8] 3Y
```

SN54LS08, SN54S08 ... FK PACKAGE
SN74LS08, SN74S08 ... FN PACKAGE
(TOP VIEW)

NC - No internal connection

FUNCTION TABLE (each gate)

INPUTS		OUTPUT
A	B	Y
H	H	H
L	X	L
X	L	L

logic diagram (each gate)

positive logic

$$Y = A \cdot B \text{ or } Y = \overline{\overline{A} + \overline{B}}$$

- Package Options Include Plastic "Small Outline" Packages, Ceramic Chip Carriers and Flat Packages, and Plastic and Ceramic DIPs

- Dependable Texas Instruments Quality and Reliability

description

These devices contain four independent 2-input OR gates.

The SN5432, SN54LS32 and SN54S32 are characterized for operation over the full military range of −55°C to 125°C. The SN7432, SN74LS32 and SN74S32 are characterized for operation from 0°C to 70°C.

FUNCTION TABLE (each gate)

INPUTS		OUTPUT
A	B	Y
H	X	H
X	H	H
L	L	L

logic symbol†

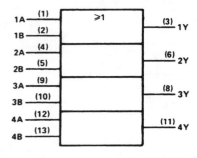

† This symbol is in accordance with ANSI/IEEE Std 91-1984 and IEC Publication 617-12.
Pin numbers shown are for D, J, N, or W packages.

**SN5432, SN54LS32, SN54S32 . . . J OR W PACKAGE
SN7432 . . . N PACKAGE
SN74LS32, SN74S32 . . . D OR N PACKAGE
(TOP VIEW)**

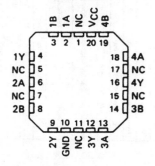

**SN54LS32, SN54S32 . . . FK PACKAGE
(TOP VIEW)**

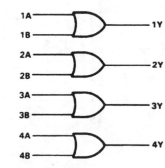

NC - No internal connection

logic diagram

positive logic

$$Y = A + B \text{ or } Y = \overline{\overline{A} \cdot \overline{B}}$$

2

TTL Devices

TEXAS INSTRUMENTS
POST OFFICE BOX 655012 • DALLAS, TEXAS 75265

2-137

Courtesy of Texas Instruments Incorporated

TYPES SN5446A, '47A, '48, '49, SN54L46, 'L47, SN54LS47, 'LS48, 'LS49, SN7446A, '47A, '48, SN74LS47, 'LS48, 'LS49 BCD-TO-SEVEN-SEGMENT DECODERS/DRIVERS

MARCH 1974 – REVISED DECEMBER 1983

'46A, '47A, 'L46, 'L47, 'LS47 feature
- Open-Collector Outputs Drive Indicators Directly
- Lamp-Test Provision
- Leading/Trailing Zero Suppression

'48, 'LS48 feature
- Internal Pull-Ups Eliminate Need for External Resistors
- Lamp-Test Provision
- Leading/Trailing Zero Suppression

'49, 'LS49 feature
- Open-Collector Outputs
- Blanking Input

- All Circuit Types Feature Lamp Intensity Modulation Capability

TYPE	ACTIVE LEVEL	DRIVER OUTPUTS OUTPUT CONFIGURATION	SINK CURRENT	MAX VOLTAGE	TYPICAL POWER DISSIPATION	PACKAGES
SN5446A	low	open-collector	40 mA	30 V	320 mW	J, W
SN5447A	low	open-collector	40 mA	15 V	320 mW	J, W
SN5448	high	2-kΩ pull-up	6.4 mA	5.5 V	265 mW	J, W
SN5449	high	open-collector	10 mA	5.5 V	165 mW	W
SN54L46	low	open-collector	20 mA	30 V	160 mW	J
SN54L47	low	open-collector	20 mA	15 V	160 mW	J
SN54LS47	low	open-collector	12 mA	15 V	35 mW	J, W
SN54LS48	high	2-kΩ pull-up	2 mA	5.5 V	125 mW	J, W
SN54LS49	high	open-collector	4 mA	5.5 V	40 mW	J, W
SN7446A	low	open-collector	40 mA	30 V	320 mW	J, N
SN7447A	low	open-collector	40 mA	15 V	320 mW	J, N
SN7448	high	2-kΩ pull-up	6.4 mA	5.5 V	265 mW	J, N
SN74LS47	low	open-collector	24 mA	15 V	35 mW	J, N
SN74LS48	high	2-kΩ pull-up	6 mA	5.5 V	125 mW	J, N
SN74LS49	high	open-collector	8 mA	5.5 V	40 mW	J, N

logic symbols

'46, '47

'48

'49

Pin numbers shown on logic notation are for D, J or N packages.

SN54L46, SN54L47 . . . J PACKAGE
SN5446A, SN5447A, SN54LS47, SN54LS48,
SN54LS48 . . . J OR W PACKAGE
SN7446A, SN7447A,
SN7448 . . . J OR N PACKAGE
SN74LS47, SN74LS48 . . . D, J OR N PACKAGE
(TOP VIEW)

SN5449 . . . W PACKAGE
SN54LS49 . . . J OR W PACKAGE
SN74LS49 . . . D, J OR N PACKAGE
(TOP VIEW)

SN54LS47, SN54LS48 . . . FK PACKAGE
SN74LS47, SN74LS48 . . . FN PACKAGE
(TOP VIEW)

SN54LS49 . . . FK PACKAGE
SN74LS49 . . . FN PACKAGE
(TOP VIEW)

NC – No internal connection

Courtesy of Texas Instruments Incorporated

TEXAS INSTRUMENTS
POST OFFICE BOX 225012 • DALLAS, TEXAS 75265

TEXAS INSTRUMENTS
POST OFFICE BOX 225012 • DALLAS, TEXAS 75265

absolute maximum ratings over operating free-air temperature range (unless otherwise noted)

Supply voltage, V_{CC} (see Note 1) .... 7 V
Input voltage .... 5.5 V
Current forced into any output in the off state .... 1 mA
Operating free-air temperature range: SN5446A, SN5447A .... −55°C to 125°C
  SN7446A, SN7447A .... 0°C to 70°C
Storage temperature range .... −65°C to 150°C

NOTE 1: Voltage values are with respect to network ground terminal.

recommended operating conditions

		SN5446A			SN5447A			SN7446A			SN7447A			UNIT
		MIN	NOM	MAX	MIN	NOM	MAX	MIN	NOM	MAX	MIN	NOM	MAX	
Supply voltage, V_{CC}		4.5	5	5.5	4.5	5	5.5	4.75	5	5.25	4.75	5	5.25	V
Off-state output voltage, $V_{O(off)}$	a thru g			30			30			15			15	V
On-state output current, $I_{O(on)}$	a thru g			40			40			40			40	mA
High-level output current, I_{OH}	BI/RBO			−200			−200			−200			−200	μA
Low-level output current, I_{OL}	BI/RBO			8			8			8			8	mA
Operating free-air temperature, T_A		−55		125	−55		125	0		70	0		70	°C

electrical characteristics over recommended operating free-air temperature range (unless otherwise noted)

	PARAMETER	TEST CONDITIONS†	MIN	TYP‡	MAX	UNIT
V_{IH}	High-level input voltage		2			V
V_{IL}	Low-level input voltage				0.8	V
V_{IK}	Input clamp voltage	V_{CC} = MIN, I_I = −12 mA			−1.5	V
V_{OH}	High-level output voltage	BI/RBO V_{CC} = MIN, V_{IH} = 2 V, V_{IL} = 0.8 V, I_{OH} = −200 μA	2.4	3.7		V
V_{OL}	Low-level output voltage	BI/RBO V_{CC} = MIN, V_{IH} = 2 V, V_{IL} = 0.8 V, I_{OL} = 8 mA		0.27	0.4	V
$I_{O(off)}$	Off-state output current	a thru g V_{CC} = MAX, V_{IH} = 2 V, V_{IL} = 0.8 V, $V_{O(off)}$ = MAX			250	μA
$V_{O(on)}$	On-state output voltage	a thru g V_{CC} = MIN, V_{IH} = 2 V, V_{IL} = 0.8 V, $I_{O(on)}$ = 40 mA		0.3	0.4	V
I_I	Input current at maximum input voltage	Any input except BI/RBO V_{CC} = MAX, V_I = 5.5 V			1	mA
I_{IH}	High-level input current	Any input except BI/RBO V_{CC} = MAX, V_I = 2.4 V			40	μA
I_{IL}	Low-level input current	Any input except BI/RBO V_{CC} = MAX, V_I = 0.4 V			−1.6	mA
I_{OS}	Short-circuit output current	BI/RBO V_{CC} = MAX SN54′ / SN74′			−4 / −4	mA
I_{CC}	Supply current	V_{CC} = MAX, See Note 2 SN54′ 64 85 / SN74′ 64 103				mA

† For conditions shown as MIN or MAX, use the appropriate value specified under recommended operating conditions.
‡ All typical values are at V_{CC} = 5 V, T_A = 25°C.
NOTE 2: I_{CC} is measured with all outputs open and all inputs at 4.5 V.

switching characteristics, V_{CC} = 5 V, T_A = 25°C

	PARAMETER	TEST CONDITIONS	MIN	TYP	MAX	UNIT
t_{off}	Turn-off time from A input	C_L = 15 pF, R_L = 120 Ω, See Note 3			100	ns
t_{on}	Turn-on time from A input				100	ns
t_{off}	Turn-off time from RBI input				100	ns
t_{on}	Turn-on time from RBI input				100	ns

NOTE 3: See General Information Section for load circuits and voltage waveforms. *t_{off} corresponds to t_{PLH} and t_{on} corresponds to t_{PHL}.

TEXAS INSTRUMENTS
POST OFFICE BOX 225012 • DALLAS, TEXAS 75265

---

description

The '46A, 'L46, '47A, '47, and 'LS47 feature active-low outputs designed for driving common-anode VLEDs or incandescent indicators directly, and the '48, '49, 'LS48, 'LS49 feature active-high outputs for driving lamp buffers or common-cathode VLEDs. All of the circuits except '49 and 'LS49 have full ripple-blanking input/output controls and a lamp test input. The '49 and 'LS49 circuits incorporate a direct blanking input. Segment identification and resultant displays are shown below. Display patterns for BCD input counts above 9 are unique symbols to authenticate input conditions.

The '46A, '47A, '48, 'L46, 'L47, 'LS47, and 'LS48 circuits incorporate automatic leading and/or trailing-edge zero-blanking control (RBI and RBO). Lamp test (LT) of these types may be performed at any time when the BI/RBO node is at a high level. All types (including the '49 and 'LS49) contain an overriding blanking input (BI) which can be used to control the lamp intensity by pulsing or to inhibit the outputs. Inputs and outputs are entirely compatible for use with TTL logic outputs.

The SN54246 through '249 and the SN54LS247/SN74LS247 through 'LS249 compose the 6 and the 9 with tails and have been designed to offer the designer a choice between two indicator fonts. The SN54249/SN74249 and SN54LS249/SN74LS249 are 16-pin versions of the 14-pin SN5449 and 'LS49. Included in the '249 circuit and 'LS249 circuits are the full functional capability for lamp test and ripple blanking, which is not available in the '49 or 'LS49 circuit.

SEGMENT IDENTIFICATION

NUMERICAL DESIGNATIONS AND RESULTANT DISPLAYS

'46A, '47A, 'L46, 'L47, 'LS47 FUNCTION TABLE

DECIMAL OR FUNCTION	INPUTS						BI/RBO†	OUTPUTS							NOTE
	LT	RBI	D	C	B	A		a	b	c	d	e	f	g	
0	H	H	L	L	L	L	H	ON	ON	ON	ON	ON	ON	OFF	1
1	H	X	L	L	L	H	H	OFF	ON	ON	OFF	OFF	OFF	OFF	
2	H	X	L	L	H	L	H	ON	ON	OFF	ON	ON	OFF	ON	
3	H	X	L	L	H	H	H	ON	ON	ON	ON	OFF	OFF	ON	
4	H	X	L	H	L	L	H	OFF	ON	ON	OFF	OFF	ON	ON	
5	H	X	L	H	L	H	H	ON	OFF	ON	ON	OFF	ON	ON	
6	H	X	L	H	H	L	H	OFF	OFF	ON	ON	ON	ON	ON	
7	H	X	L	H	H	H	H	ON	ON	ON	OFF	OFF	OFF	OFF	
8	H	X	H	L	L	L	H	ON	ON	ON	ON	ON	ON	ON	
9	H	X	H	L	L	H	H	ON	ON	ON	OFF	OFF	ON	ON	
10	H	X	H	L	H	L	H	OFF	OFF	OFF	ON	ON	OFF	ON	
11	H	X	H	L	H	H	H	OFF	OFF	ON	ON	OFF	OFF	ON	
12	H	X	H	H	L	L	H	OFF	ON	OFF	OFF	OFF	ON	ON	
13	H	X	H	H	L	H	H	ON	OFF	OFF	ON	OFF	ON	ON	
14	H	X	H	H	H	L	H	OFF	OFF	OFF	ON	ON	ON	ON	
15	H	X	H	H	H	H	H	OFF	OFF	OFF	OFF	OFF	OFF	OFF	
BI	X	X	X	X	X	X	L	OFF	OFF	OFF	OFF	OFF	OFF	OFF	2
RBI	H	L	L	L	L	L	L	OFF	OFF	OFF	OFF	OFF	OFF	OFF	3
LT	L	X	X	X	X	X	H	ON	ON	ON	ON	ON	ON	ON	4

H = high level, L = low level, X = irrelevant

NOTES: 1. The blanking input (BI) must be open or held at a high logic level when output functions 0 through 15 are desired. The ripple-blanking input (RBI) must be open or high if blanking of a decimal zero is not desired.
2. When a low logic level is applied directly to the blanking input (BI), all segment outputs are off regardless of the level of any other input.
3. When ripple-blanking input (RBI) and inputs A, B, C, and D are at a low level with the lamp test input high, all segment outputs go off and the ripple-blanking output (RBO) goes to a low level (response condition).
4. When the blanking input/ripple blanking output (BI/RBO) is open or held high and a low is applied to the lamp test input, all segment outputs are on.

† BI/RBO is wire AND logic serving as blanking input (BI) and/or ripple blanking output (RBO).

TEXAS INSTRUMENTS
POST OFFICE BOX 225012 • DALLAS, TEXAS 75265

234

Recommended Operating Conditions

Sym	Parameter		DM5474			DM7474			Units
			Min	Nom	Max	Min	Nom	Max	
V_{CC}	Supply Voltage		4.5	5	5.5	4.75	5	5.25	V
V_{IH}	High Level Input Voltage		2			2			V
V_{IL}	Low Level Input Voltage				0.8			0.8	V
I_{OH}	High Level Output Current				−0.4			−0.4	mA
I_{OL}	Low Level Output Current				16			16	mA
f_{CLK}	Clock Frequency		0		20	0		20	MHz
t_W	Pulse Width	Clock High	30			30			ns
		Clock Low	37			37			
		Clear Low	30			30			
		Preset Low	30			30			
t_{su}	Input Setup Time (Note 1)		20↑			20↑			ns
t_H	Input Hold Time (Note 1)		5↑			5↑			ns
T_A	Free Air Operating Temperature		−55		125	0		70	°C

Note 1: The symbol (↑) indicates the rising edge of the clock pulse is used for reference.

National Semiconductor

DM5474/DM7474 Dual Positive-Edge-Triggered D Flip-Flops with Preset, Clear and Complementary Outputs

General Description

This device contains two independent positive-edge-triggered D flip-flops with complementary outputs. The information on the D input is accepted by the flip-flops on the positive going edge of the clock pulse. The triggering occurs at a voltage level and is not directly related to the transition time of the rising edge of the clock. The data on the D input may be changed while the clock is low or high without affecting the outputs as long as the data setup and hold times are not violated. A low logic level on the preset or clear inputs will set or reset the outputs regardless of the logic levels of the other inputs.

Absolute Maximum Ratings (Note 1)

Supply Voltage	7V
Input Voltage	5.5V
Storage Temperature Range	−65°C to 150°C

Note 1: The "Absolute Maximum Ratings" are those values beyond which the safety of the device can not be guaranteed. The device should not be operated at these limits. The parametric values defined in the "Electrical Characteristics" table are not guaranteed at the absolute maximum ratings. The "Recommended Operating Conditions" table will define the conditions for actual device operation.

Connection Diagram

Dual-In-Line Package

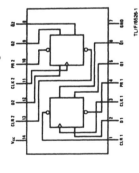

TL/F/6526-1

DM5474 (J)   DM7474 (N)

Function Table

Inputs				Outputs	
PR	CLR	CLK	D	Q	Q̄
L	H	X	X	H	L
H	L	X	X	L	H*
L	L	X	X	H*	H*
H	H	↑	H	H	L
H	H	↑	L	L	H
H	H	L	X	Q₀	Q̄₀

H = High Logic Level
X = Either Low or High Logic Level
L = Low Logic Level
↑ = Positive-going transition of the clock.
* = This configuration is nonstable; that is, it will not persist when either the preset and/or clear inputs return to their inactive (high) level.
Q_0 = The output logic level of Q before the indicated input conditions were established.

TYPES SN5476, SN54H76, SN54LS76A, SN7476, SN74H76, SN74LS76A
DUAL J-K FLIP-FLOPS WITH PRESET AND CLEAR

SN5476, SN54H76, SN54LS76A . . . J OR W PACKAGE
SN7476, SN74H76 . . . J OR N PACKAGE
SN74LS76A . . . D, J OR N PACKAGE

REVISED DECEMBER 1983

- **Package Options Include Plastic and Ceramic DIPs**
- **Dependable Texas Instruments Quality and Reliability**

(TOP VIEW)

1CLK	1 ⌐16	1K
1PRE	2 15	1Q
1CLR	3 14	1Q̄
1J	4 13	GND
VCC	5 12	2K
2CLK	6 11	2Q
2PRE	7 10	2Q̄
2CLR	8 9	2J

description

The '76 and 'H76 contain two independent J-K flip-flops with individual J-K, clock, preset, and clear inputs. The '76 and 'H76 are positive-edge-triggered flip-flops. J-K input is loaded into the master while the clock is high and transferred to the slave on the high-to-low transition. For these devices the J and K inputs must be stable while the clock is high.

The 'LS76A contain two independent negative-edge-triggered flip-flops. The J and K inputs must be stable one setup time prior to the high-to-low clock transition for predictable operation. The preset and clear are asynchronous active low inputs. When low they override the clock and data inputs forcing the outputs to the steady state levels as shown in the function table.

The SN5476, SN54H76, and the SN54LS76A are characterized for operation over the full military temperature range of −55°C to 125°C. The SN7476, SN74H76, and the SN74LS76A are characterized for operation from 0°C to 70°C.

'76, 'H76 FUNCTION TABLE

	INPUTS				OUTPUTS	
PRE	CLR	CLK	J	K	Q	Q̄
L	H	X	X	X	H	L
H	L	X	X	X	L	H
L	L	X	X	X	H†	H†
H	H	⊓	L	L	Q₀	Q̄₀
H	H	⊓	H	L	H	L
H	H	⊓	L	H	L	H
H	H	⊓	H	H		TOGGLE

'LS76A FUNCTION TABLE

	INPUTS				OUTPUTS	
PRE	CLR	CLK	J	K	Q	Q̄
L	H	X	X	X	H	L
H	L	X	X	X	L	H
L	L	X	X	X	H†	H†
H	H	↓	L	L	Q₀	Q̄₀
H	H	↓	H	L	H	L
H	H	↓	L	H	L	H
H	H	↓	H	H		TOGGLE
H	H	H	X	X	Q₀	Q̄₀

† This configuration is nonstable; that is, it will not persist when either preset or clear returns to its inactive (high) level.

FOR CHIP CARRIER INFORMATION, CONTACT THE FACTORY

TEXAS INSTRUMENTS
POST OFFICE BOX 225012 • DALLAS, TEXAS 75265

236

TYPES SN5476, SN54H76, SN7476, SN74H76 DUAL J-K FLIP-FLOPS WITH PRESET AND CLEAR

logic diagrams

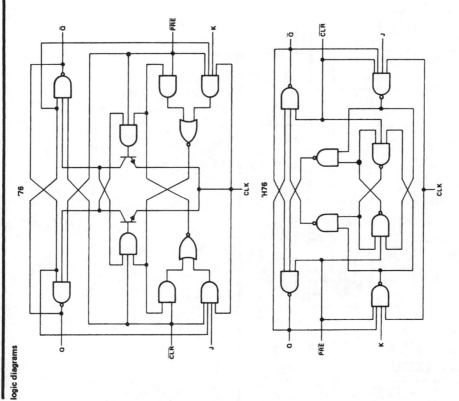

'76

'H76

TEXAS INSTRUMENTS
POST OFFICE BOX 225012 • DALLAS, TEXAS 75265

recommended operating conditions

			SN5476			SN7476			UNIT
			MIN	NOM	MAX	MIN	NOM	MAX	
V_{CC}	Supply voltage		4.5	5	5.5	4.75	5	5.25	V
V_{IH}	High-level input voltage		2			2			V
V_{IL}	Low-level input voltage				0.8			0.8	V
I_{OH}	High-level output current				−0.4			−0.4	mA
I_{OL}	Low-level output current				16			16	mA
t_w	Pulse duration	CLK high	20			20			ns
		CLK low	47			47			
		PRE or CLR low	25			25			
t_{su}	Input setup time before CLK ↑		0			0			ns
t_h	Input hold time-data after CLK ↓		0			0			ns
T_A	Operating free-air temperature		−55		125	0		70	°C

electrical characteristics over recommended operating free-air temperature range (unless otherwise noted)

PARAMETER	TEST CONDITIONS†			SN5476			SN7476			UNIT
				MIN	TYP	MAX	MIN	TYP	MAX	
V_{IK}	V_{CC} = MIN,	I_I = −12 mA				−1.5			−1.5	V
V_{OH}	V_{CC} = MIN,	V_{IH} = 2 V,	V_{IL} = 0.8 V,	2.4	3.4		2.4	3.4		V
	I_{OH} = −0.4 mA									
V_{OL}	V_{CC} = MIN,	V_{IH} = 2 V,	V_{IL} = 0.8 V,		0.2	0.4		0.2	0.4	V
	I_{OL} = 16 mA									
I_I	V_{CC} = MAX,	V_I = 5.5 V				1			1	mA
I_{IH}	J or K	V_{CC} = MAX,	V_I = 2.4 V			40			40	µA
	All other					80			80	
I_{IL}	J or K	V_{CC} = MAX,	V_I = 0.4 V			−1.6			−1.6	mA
	All other‡					−3.2			−3.2	
I_{OS}§	V_{CC} = MAX	See Note 2		−20		−57	−18		−57	mA
I_{CC}	V_{CC} = MAX,				10	20		10	20	mA

† For conditions shown as MIN or MAX, use the appropriate value specified under recommended operating conditions.
‡ All typical values are at V_{CC} = 5 V, T_A = 25°C.
§ Not more than one output should be shorted at a time.
*Clear is tested with preset high and preset is tested with clear high.
NOTE 2: With all outputs open, I_{CC} is measured with the Q and Q̄ outputs high in turn. At the time of measurement, the clock input is grounded.

switching characteristics, V_{CC} = 5 V, T_A = 25°C (see note 3)

PARAMETER	FROM (INPUT)	TO (OUTPUT)	TEST CONDITIONS		MIN	TYP	MAX	UNIT
f_{max}			R_L = 400 Ω,	C_L = 15 pF	15	20		MHz
t_{PLH}	PRE or CLR	Q or Q̄				16	25	ns
t_{PHL}		Q or Q̄				25	40	ns
t_{PLH}	CLK	Q or Q̄				16	25	ns
t_{PHL}		Q or Q̄				25	40	ns

NOTE 3: See General Information Section for load circuits and voltage waveforms.

TEXAS
INSTRUMENTS
POST OFFICE BOX 225012 • DALLAS, TEXAS 75265

---

logic diagrams (continued)

'LS76A

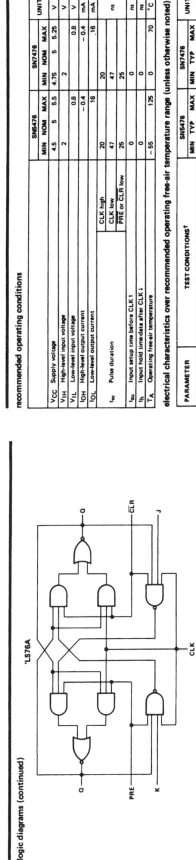

logic symbols

76, 'H76

'LS76A

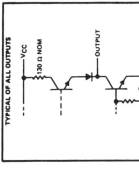

Pin numbers shown on logic notation are for D, J or N packages.

76

schematics of inputs and outputs

EQUIVALENT OF EACH INPUT

I_{IL} MAX	R_{eq} NOM
−1.6 mA	4 kΩ
−3.2 mA	2 kΩ

TYPICAL OF ALL OUTPUTS

TEXAS
INSTRUMENTS
POST OFFICE BOX 225012 • DALLAS, TEXAS 75265

TYPES SN5483A, SN54LS83A, SN7483A, SN74LS83A
4-BIT BINARY FULL ADDERS WITH FAST CARRY

MARCH 1974—REVISED DECEMBER 1983

- **Full-Carry Look-Ahead across the Four Bits**

- **Systems Achieve Partial Look-Ahead Performance with the Economy of Ripple Carry**

- **SN54283/SN74283 and SN54LS283/SN74LS283 Are Recommended For New Designs as They Feature Supply Voltage and Ground on Corner Pins to Simplify Board Layout**

TYPE	TYPICAL ADD TIMES		TYPICAL POWER DISSIPATION PER 4-BIT ADDER
	TWO 8-BIT WORDS	TWO 16-BIT WORDS	
'83A	23 ns	43 ns	310 mW
'LS83A	25 ns	45 ns	95 mW

description

These improved full adders perform the addition of two 4-bit binary numbers. The sum (Σ) outputs are provided for each bit and the resultant carry (C4) is obtained from the fourth bit. These adders feature full internal look ahead across all four bits generating the carry term in ten nanoseconds typically. This provides the system designer with partial look-ahead performance at the economy and reduced package count of a ripple-carry implementation.

The adder logic, including the carry, is implemented in its true form meaning that the end-around carry can be accomplished without the need for logic or level inversion.

Designed for medium-speed applications, the circuits utilize transistor-transistor logic that is compatible with most other TTL families and other saturated low-level logic families.

Series 54 and 54LS circuits are characterized for operation over the full military temperature range of −55°C to 125°C, and Series 74 and 74LS circuits are characterized for operation from 0°C to 70°C.

SN5483A, SN54LS83A . . . J OR W PACKAGE
SN7483A . . . J OR N PACKAGE
SN74LS83A . . . D, J OR N PACKAGE
(TOP VIEW)

A4	1	16	B4
Σ3	2	15	Σ4
A3	3	14	C4
B3	4	13	C0
V$_{CC}$	5	12	GND
Σ2	6	11	B1
B2	7	10	A1
A2	8	9	Σ1

SN54LS83A . . . FK PACKAGE
SN74LS83A . . . FN PACKAGE
(TOP VIEW)

NC - No internal connection

FUNCTION TABLE

H = high level, L = low level

NOTE: Input conditions at A1, B1, A2, B2, and C0 are used to determine outputs Σ1 and Σ2 and the value of the internal carry C2. The values at C2, A3, B3, A4, and B4 are then used to determine outputs Σ3, Σ4, and C4.

logic diagram

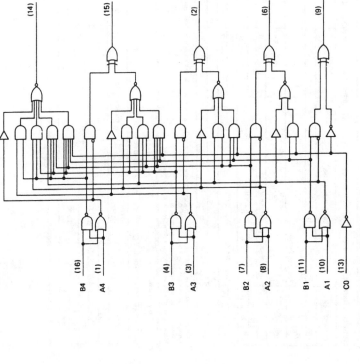

Pin numbers shown on logic notation are for D, J or N packages.

absolute maximum ratings over operating free-air temperature range (unless otherwise noted)

Supply voltage, V$_{CC}$ (see Note 1)		7 V
Input voltage: '83A		5.5 V
'LS83A		7 V
Interemitter voltage (see Note 2)		5.5 V
Operating free-air temperature range: SN5483A, SN54LS83A		−55°C to 125°C
SN7483A, SN74LS83A		0°C to 70°C
Storage temperature range		−65°C to 150°C

NOTES: 1. Voltage values, except interemitter voltage, are with respect to network ground terminal.
2. This is the voltage between two emitters of a multiple-emitter transistor. This rating applies for the '83A only between the following pairs: A1 and B1, A2 and B2, A3 and B3, A4 and B4.

Recommended Operating Conditions

Sym	Parameter	DM5485			DM7485			Units
		Min	Nom	Max	Min	Nom	Max	
V_{CC}	Supply Voltage	4.5	5	5.5	4.75	5	5.25	V
V_{IH}	High Level Input Voltage	2			2			V
V_{IL}	Low Level Input Voltage			0.8			0.8	V
I_{OH}	High Level Output Current			-0.8			-0.8	mA
I_{OL}	Low Level Output Current			16			16	mA
T_A	Free Air Operating Temperature	-55		125	0		70	°C

Electrical Characteristics over recommended operating free air temperature (unless otherwise noted)

Sym	Parameter	Conditions		Min	Typ (Note 1)	Max	Units
V_I	Input Clamp Voltage	V_{CC} = Min, I_I = -12 mA				-1.5	V
V_{OH}	High Level Output Voltage	V_{CC} = Min, I_{OH} = Max, V_{IL} = Max, V_{IH} = Min		2.4			V
V_{OL}	Low Level Output Voltage	V_{CC} = Min, I_{OL} = Max, V_{IH} = Min, V_{IL} = Max				0.4	V
I_I	Input Current @ Max Input Voltage	V_{CC} = Max, V_I = 5.5V				1	mA
I_{IH}	High Level Input Current	V_{CC} = Max, V_I = 2.4V	A<B			40	μA
			A>B			40	
			Others			120	
I_{IL}	Low Level Input Current	V_{CC} = Max, V_I = 0.4V	A<B			-1.6	mA
			A>B			-1.6	
			Others			-4.8	
I_{OS}	Short Circuit Output Current	V_{CC} = Max (Note 2)	DM54	-20		-55	mA
			DM74	-18		-55	
I_{CC}	Supply Current	V_{CC} = Max (Note 3)			55	88	mA

Note 1: All typicals are at V_{CC}=5V, T_A=25°C.
Note 2: Not more than one output should be shorted at a time.
Note 3: I_{CC} is measured with all outputs open, A=B input grounded and all other inputs at 4.5V.

National Semiconductor

DM5485/DM7485 4-Bit Magnitude Comparators

General Description

These 4-bit magnitude comparators perform comparison of straight binary or BCD codes. Three fully-decoded decisions about two 4-bit words (A, B) are made and are externally available at three outputs. These devices are fully expandable to any number of bits without external gates. Words of greater length may be compared by connecting comparators in cascade. The A>B, A<B, and A=B outputs of a stage handling less-significant bits are connected to the corresponding inputs of the next stage handling more-significant bits. The stage handling the least-significant bits must have a high-level voltage applied to the A=B input. The cascading paths are implemented with only a two-gate-level delay to reduce overall comparison times for long words.

Features

- Typical power dissipation 275 mW
- Typical delay (4-bit words) 23 ns

Absolute Maximum Ratings (Note 1)

Supply Voltage   7V
Input Voltage   5.5V
Storage Temperature Range   -65°C to 150°C

Note 1: The "Absolute Maximum Ratings" are those values beyond which the safety of the device can not be guaranteed. The device should not be operated at these limits. The parametric values defined in the "Electrical Characteristics" table are not guaranteed at the absolute maximum ratings. The "Recommended Operating Conditions" table will define the conditions for actual device operation.

Connection Diagram

Dual-In-Line Package

TL/F/6530-1

5485 (J)
7485 (N)

- Package Options Include Plastic "Small Outline" Packages, Ceramic Chip Carriers and Flat Packages, and Standard Plastic and Ceramic 300-mil DIPs

- Dependable Texas Instruments Quality and Reliability

TYPE	TYPICAL AVERAGE PROPAGATION DELAY TIME	TYPICAL TOTAL POWER DISSIPATION
'86	14 ns	150 mW
'LS86A	10 ns	30.5 mW
'S86	7 ns	250 mW

description

These devices contain four independent 2-input Exclusive-OR gates. They perform the Boolean functions $Y = A \oplus B = \bar{A}B + A\bar{B}$ in positive logic.

A common application is as a true/complement element. If one of the inputs is low, the other input will be reproduced in true form at the output. If one of the inputs is high, the signal on the other input will be reproduced inverted at the output.

The SN5486, 54LS86A, and the SN54S86 are characterized for operation over the full military temperature range of −55°C to 125°C. The SN7486, SN74LS86A, and the SN74S86 are characterized for operation from 0°C to 70°C.

SN5486, SN54LS86A, SN54S86 . . . J OR W PACKAGE
SN7486 . . . N PACKAGE
SN74LS86A, SN74S86 . . . D OR N PACKAGE
(TOP VIEW)

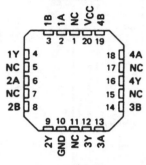

SN54LS86A, SN54S86 . . . FK PACKAGE
(TOP VIEW)

NC - No internal connection

TTL Devices

exclusive-OR logic

An exclusive-OR gate has many applications, some of which can be represented better by alternative logic symbols.

EXCLUSIVE-OR

These are five equivalent Exclusive-OR symbols valid for an '86 or 'LS86A gate in positive logic; negation may be shown at any two ports.

LOGIC IDENTITY ELEMENT	EVEN-PARITY	ODD-PARITY ELEMENT

The output is active (low) if all inputs stand at the same logic level (i.e., A = B).

The output is active (low) if an even number of inputs (i.e., 0 or 2) are active.

The output is active (high) if an odd number of inputs (i.e., only 1 of the 2) are active.

PRODUCTION DATA documents contain information current as of publication date. Products conform to specifications per the terms of Texas Instruments standard warranty. Production processing does not necessarily include testing of all parameters.

TEXAS INSTRUMENTS
POST OFFICE BOX 655012 • DALLAS, TEXAS 75265

2-271

Courtesy of Texas Instruments Incorporated

240

TYPES SN5490A, '92A, '93A, SN54L90, 'L93, SN54LS90, 'LS92, 'LS93, SN7490A, '92A, '93A, SN74LS90, 'LS92, 'LS93 DECADE, DIVIDE-BY-TWELVE, AND BINARY COUNTERS

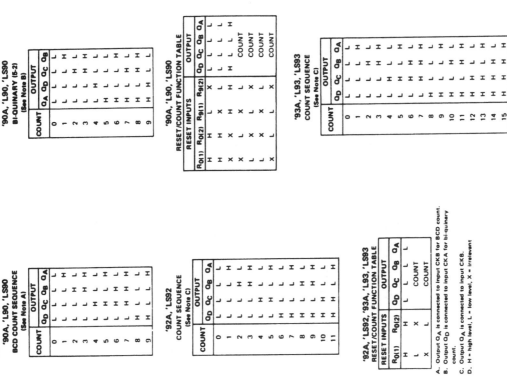

'90A, 'L90, 'LS90 BCD COUNT SEQUENCE (See Note A)

COUNT	OUTPUT			
	QD	QC	QB	QA
0	L	L	L	L
1	L	L	L	H
2	L	L	H	L
3	L	L	H	H
4	L	H	L	L
5	L	H	L	H
6	L	H	H	L
7	L	H	H	H
8	H	L	L	L
9	H	L	L	H

'90A, 'L90, 'LS90 BI-QUINARY (5-2) (See Note B)

COUNT	OUTPUT			
	QA	QD	QC	QB
0	L	L	L	L
1	L	L	L	H
2	L	L	H	L
3	L	L	H	H
4	L	H	L	L
5	H	L	L	L
6	H	L	L	H
7	H	L	H	L
8	H	L	H	H
9	H	H	L	L

'90A, 'L90, 'LS90 RESET/COUNT FUNCTION TABLE

RESET INPUTS				OUTPUT			
R0(1)	R0(2)	R9(1)	R9(2)	QD	QC	QB	QA
H	H	L	X	L	L	L	L
H	H	X	L	L	L	L	L
X	X	H	H	H	L	L	H
X	L	X	L	COUNT			
L	X	L	X	COUNT			
L	X	X	L	COUNT			
X	L	L	X	COUNT			

'93A, 'L93, 'LS93 COUNT SEQUENCE (See Note C)

COUNT	OUTPUT			
	QD	QC	QB	QA
0	L	L	L	L
1	L	L	L	H
2	L	L	H	L
3	L	L	H	H
4	L	H	L	L
5	L	H	L	H
6	L	H	H	L
7	L	H	H	H
8	H	L	L	L
9	H	L	L	H
10	H	L	H	L
11	H	L	H	H
12	H	H	L	L
13	H	H	L	H
14	H	H	H	L
15	H	H	H	H

'92A, 'LS92 COUNT SEQUENCE (See Note C)

COUNT	OUTPUT			
	QD	QC	QB	QA
0	L	L	L	L
1	L	L	L	H
2	L	L	H	L
3	L	L	H	H
4	L	H	L	L
5	L	H	L	H
6	H	L	L	L
7	H	L	L	H
8	H	L	H	L
9	H	L	H	H
10	H	H	L	L
11	H	H	L	H

'92A, 'LS92, '93A, 'L93, 'LS93 RESET/COUNT FUNCTION TABLE

RESET INPUTS		OUTPUT			
R0(1)	R0(2)	QD	QC	QB	QA
H	H	L	L	L	L
L	X	COUNT			
X	L	COUNT			

NOTES: A. Output QA is connected to input CKB for BCD count.
B. Output QD is connected to input CKA for bi-quinary count.
C. Output QA is connected to input CKB.
D. H = high level, L = low level, X = irrelevant

MARCH 1974 — REVISED DECEMBER 1983

TYPES SN5490A, SN5492A, SN5493A, SN54L90, SN54L93, SN54LS90, SN54LS92, SN54LS93, SN7490A, SN7492A, SN7493A, SN74LS90, SN74LS92, SN74LS93 DECADE, DIVIDE-BY-TWELVE, AND BINARY COUNTERS

- '90A, 'L90, 'LS90 ... DECADE COUNTERS
- '92A, 'LS92 ... DIVIDE-BY-TWELVE COUNTERS
- '93A, 'L93, 'LS93 ... 4-BIT BINARY COUNTERS

TYPES	TYPICAL POWER DISSIPATION
'90A	145 mW
'L90	20 mW
'LS90	45 mW
'92A, '93A	130 mW
'LS92, 'LS93	45 mW
'L93	16 mW

description

Each of these monolithic counters contains four master-slave flip-flops and additional gating to provide a divide-by-two counter and a three-stage binary counter for which the count cycle length is divide-by-five for the '90A, 'L90, and 'LS90, divide-by-six for the '92A and 'LS92, and divide-by-eight for the '93A, 'L93, and 'LS93.

All of these counters have a gated zero reset and the '90A, 'L90, and 'LS90 also have gated set-to-nine inputs for use in BCD nine's complement applications.

To use their maximum count length (decade, divide-by-twelve, or four-bit binary) of these counters, the CKB input is connected to the QA output. The input count pulses are applied to CKA input and the outputs are as described in the appropriate function table. A symmetrical divide-by-ten count can be obtained from the '90A, 'L90, or 'LS90 counters by connecting the QD output to the CKA input and applying the input count to the CKB input which gives a divide-by-ten square wave at output QA.

SN5490A, SN54LS90 ... J PACKAGE
SN54L90 ... J PACKAGE
SN7490A ... J OR N PACKAGE
SN74LS90 ... D, J OR N PACKAGE
(TOP VIEW)

```
         ___ ___
CKB   [1]       [14] CKA
R0(1) [2]       [13] NC
R0(2) [3]       [12] QA
NC    [4]       [11] QD
VCC   [5]       [10] GND
R9(1) [6]       [9]  QB
R9(2) [7]       [8]  QC
```

SN5492A, SN54LS92 ... J OR W PACKAGE
SN7492A ... J OR N PACKAGE
SN74LS92 ... D, J OR N PACKAGE
(TOP VIEW)

```
         ___ ___
CKB   [1]       [14] CKA
NC    [2]       [13] NC
NC    [3]       [12] QA
NC    [4]       [11] QB
VCC   [5]       [10] QC
R0(1) [6]       [9]  QD
R0(2) [7]       [8]  NC
```

SN5493A, SN54LS93 ... J OR W PACKAGE
SN7493A ... J OR N PACKAGE
SN74LS93 ... D, J OR N PACKAGE
(TOP VIEW)

```
         ___ ___
CKB   [1]       [14] CKA
R0(1) [2]       [13] NC
R0(2) [3]       [12] QD
NC    [4]       [11] QC
VCC   [5]       [10] QB
NC    [6]       [9]  QA
NC    [7]       [8]  GND
```

SN54L93 ... J PACKAGE
(TOP VIEW)

```
         ___ ___
R0(1) [1]       [14] CKA
R0(2) [2]       [13] QD
NC    [3]       [12] QC
VCC   [4]       [11] GND
NC    [5]       [10] QB
NC    [6]       [9]  QA
NC    [7]       [8]  CKB
```

NC - No internal connection

For new chip carrier design, use 'LS290, 'LS292, and 'LS293.

TEXAS INSTRUMENTS
POST OFFICE BOX 225012 • DALLAS, TEXAS 75265

TYPES SN54121, SN54L121, SN74121 MONOSTABLE MULTIVIBRATORS WITH SCHMITT-TRIGGER INPUTS

REVISED MAY 1983

- Programmable Output Pulse Width
  With R_{int} . . . 35 ns Typ
  With R_{ext}/C_{ext} . . . 40 ns to 28 Seconds
- Internal Compensation for Virtual Temperature Independence
- Jitter-Free Operation up to 90% Duty Cycle
- Inhibit Capability

SN54121 . . . J OR W PACKAGE
SN54L121 . . . J PACKAGE
SN74121 . . . J OR N PACKAGE

(TOP VIEW)

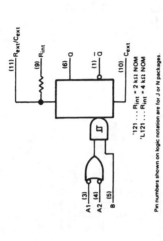

NC – No internal connection.

FUNCTION TABLE

INPUTS			OUTPUTS	
A1	A2	B	Q	Q̄
L	X	H	L	H
X	L	H	L	H
X	X	L	L	H
H	H	X	L	H
H	↓	H	⊓	⊔
↓	H	H	⊓	⊔
↓	↓	H	⊓	⊔
L	X	↑	⊓	⊔
X	L	↑	⊓	⊔

For explanation of function table symbols, see page
† These lines of the function table assume that the indicated steady state conditions at the A and B inputs have been setup long enough to complete any pulse started before the setup.

description

These multivibrators feature dual negative-transition-triggered inputs and a single positive-transition-triggered input which can be used as an inhibit input. Complementary output pulses are provided.

Pulse triggering occurs at a particular voltage level and is not directly related to the transition time of the input pulse. Schmitt-trigger input circuitry (TTL hysteresis) for the B input allows jitter-free triggering from inputs with transition rates as slow as 1 volt/second, providing the circuit with an excellent noise immunity of typically 1.2 volts. A high immunity to VCC noise of typically 1.5 volts is also provided by internal latching circuitry.

Once fired, the outputs are independent of further transitions of the inputs and are a function only of the timing components. Input pulses may be of any duration relative to the output pulse. Output pulse length may be varied from 40 nanoseconds to 28 seconds by choosing appropriate timing components. With no external timing components (i.e., R_{int} connected to V_{CC}, C_{ext} and R_{ext}/C_{ext} open), an output pulse of typically 30 or 35 nanoseconds is achieved which may be used as a d-c triggered reset signal. Output rise and fall times are TTL compatible and independent of pulse length.

Pulse width stability is achieved through internal compensation and is virtually independent of VCC and temperature. In most applications, pulse stability will only be limited by the accuracy of external timing components.

Jitter-free operation is maintained over the full temperature and VCC ranges for more than six decades of timing capacitance (10 pF to 10 μF) and more than one decade of timing resistance (2 kΩ to 30 kΩ for the SN54121/SN54L121 and 2 kΩ to 40 kΩ for the SN74121). Throughout these ranges, pulse width is defined by the relationship $t_{W(out)} = C_{ext}R_T\ln2 \approx 0.7 C_{ext}R_T$. In circuits where pulse cutoff is not critical, timing capacitance up to 1000 μF and timing resistance as low as 1.4 kΩ may be used. Also, the range of jitter-free output pulse widths is extended if VCC is held to 5 volts and free-air temperature is 25°C. Duty cycles as high as 90% are achieved when using maximum recommended R_T. Higher duty cycles are available if a certain amount of pulse-width jitter is allowed.

logic diagram (positive logic)

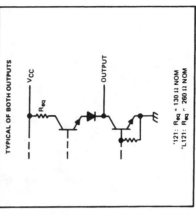

'121 . . . $R_{int} \approx 2$ kΩ NOM
'L121 . . . $R_{int} \approx 4$ kΩ NOM

Pin numbers shown on logic notation are for J or N packages.

NOTES: 1. An external capacitor may be connected between C_{ext} (positive) and R_{ext}/C_{ext}.
2. To use the internal timing resistor, connect R_{int} to V_{CC}. For improved pulse width accuracy and repeatability, connect an external resistor between R_{ext}/C_{ext} and V_{CC} with R_{int} open-circuited.

schematics of inputs and outputs

EQUIVALENT OF EACH INPUT

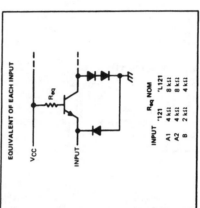

INPUT	R_{eq} NOM '121	'L121
A1	4 kΩ	8 kΩ
A2	4 kΩ	8 kΩ
B	2 kΩ	4 kΩ

TYPICAL OF BOTH OUTPUTS

'121: $R_{eq} \approx 130$ Ω NOM
'L121: $R_{eq} \approx 260$ Ω NOM

PRODUCTION DATA
This document contains information current as of publication date. Products conform to specifications per the terms of Texas Instruments standard warranty. Production processing does not necessarily include testing of all parameters.

TEXAS INSTRUMENTS
POST OFFICE BOX 225012 • DALLAS, TEXAS 75265

TEXAS INSTRUMENTS
POST OFFICE BOX 225012 • DALLAS, TEXAS 75265

Courtesy of Texas Instruments Incorporated

TYPES SN54LS139A, SN54S139, SN74LS139A, SN74S139
DUAL 2-LINE TO 4-LINE DECODERS/DEMULTIPLEXERS

REVISED APRIL 1985

- **Designed Specifically for High-Speed:**
  Memory Decoders
  Data Transmission Systems
- **Two Fully Independent 2-to-4-Line Decoders/Demultiplexers**
- **Schottky Clamped for High Performance**

description

These Schottky-clamped TTL MSI circuits are designed to be used in high-performance memory systems or data-routing applications requiring very short propagation delay times. In high-performance memory systems these decoders can be used to minimize the effects of system decoding. When employed with high-speed memories utilizing a fast enable circuit the delay times of these decoders and the enable circuit of the memory are usually less than the typical access time of the memory. This means that the effective system delay introduced by the Schottky-clamped system decoder is negligible.

The circuit comprises two individual two-line to four-line decoders in a single package. The active-low enable input can be used as a data line in demultiplexing applications.

All of these decoders/demultiplexers feature fully buffered inputs, each of which represents only one normalized load to its driving circuit. All inputs are clamped with high-performance Schottky diodes to suppress line-ringing and to simplify system design. The SN54LS139A and SN54S139 are characterized for operation over range of −55°C to 125°C. The SN74LS139A and SN74S139 are characterized for operation from 0°C to 70°C.

SN54LS139A, SN54S139 . . . J OR W PACKAGE
SN74LS139A, SN74S139 . . . D, J OR N PACKAGE
(TOP VIEW)

1G̅	1	16 V_CC
1A	2	15 2G
1B	3	14 2A
1Y0	4	13 2B
1Y1	5	12 2Y0
1Y2	6	11 2Y1
1Y3	7	10 2Y2
GND	8	9 2Y3

SN54LS139A, SN54S139 . . . FK PACKAGE
SN74LS139A, SN74S139 . . . FN PACKAGE
(TOP VIEW)

NC − No internal connection

FUNCTION TABLE

INPUTS			OUTPUTS			
ENABLE	SELECT					
G̅	B	A	Y0	Y1	Y2	Y3
H	X	X	H	H	H	H
L	L	L	L	H	H	H
L	L	H	H	L	H	H
L	H	L	H	H	L	H
L	H	H	H	H	H	L

H = high level, L = low level, X = irrelevant

logic diagram

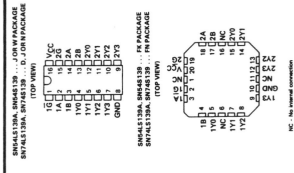

ENABLE 1G̅ (1)

SELECT
INPUTS { 1A (2), 1B (3)

ENABLE 2G̅ (15)

SELECT
INPUTS { 2A (14), 2B (13)

DATA OUTPUTS: 1Y0 (4), 1Y1 (5), 1Y2 (6), 1Y3 (7), 2Y0 (12), 2Y1 (11), 2Y2 (10), 2Y3 (9)

Pin numbers shown on logic notation are for D, J or N packages.

schematics of inputs and outputs

EQUIVALENT OF EACH INPUT OF 'LS139A

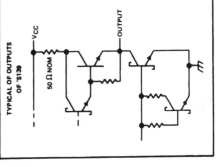

20 kΩ NOM

EQUIVALENT OF EACH INPUT OF 'S139

2.8 kΩ NOM

TYPICAL OF OUTPUTS OF 'LS139A

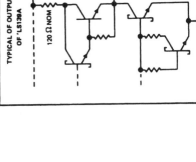

120 Ω NOM

TYPICAL OF OUTPUTS OF 'S139

50 Ω NOM

absolute maximum ratings over operating free-air temperature range (unless otherwise noted)

Supply voltage, V_CC (see Note 1)	7 V
Input voltage: 'LS139A, 'LS139	7 V
'S139	5.5 V
Operating free-air temperature range: SN54LS139A, SN54S139	−55°C to 125°C
SN74LS139A, SN54S139	0°C to 70°C
Storage temperature range	−65°C to 150°C

NOTE 1: Voltage values are with respect to network ground terminal.

TEXAS INSTRUMENTS
POST OFFICE BOX 225012 • DALLAS, TEXAS 75265

TEXAS INSTRUMENTS
POST OFFICE BOX 225012 • DALLAS, TEXAS 75265

TYPES SN54S139, SN74S139
DUAL 2-LINE TO 4-LINE DECODERS/DEMULTIPLEXERS

recommended operating conditions

	SN54S139 MIN	NOM	MAX	SN74S139 MIN	NOM	MAX	UNIT
V_{CC} Supply voltage	4.5	5	5.5	4.75	5	5.25	V
V_{IH} High-level input voltage	2			2			V
V_{IL} Low-level input voltage			0.8			0.8	V
I_{OH} High-level output current			-1			-1	mA
I_{OL} Low-level output current			20			20	mA
T_A Operating free-air temperature	-55		125	0		70	°C

electrical characteristics over recommended operating free-air temperature range (unless otherwise noted)

PARAMETER	TEST CONDITIONS†		SN54S138 SN74S138‡ MIN	TYP‡	MAX	UNIT
V_{IK}	V_{CC} = MIN,	I_I = 18 mA			-1.2	V
V_{OH}	V_{CC} = MIN, V_{IH} = 2 V, V_{IL} = 0.8 V, I_{OH} = -1 mA	SN54S'	2.5	3.4		V
		SN74S'	2.7	3.4		
V_{OL}	V_{CC} = MIN, V_{IH} = 2 V, V_{IL} = 0.8 V, I_{OL} = 20 mA				0.5	V
I_I	V_{CC} = MAX, V_I = 5.5 V				1	mA
I_{IH}	V_{CC} = MAX, V_I = 2.7 V				50	µA
I_{IL}	V_{CC} = MAX, V_I = 0.5 V				-2	mA
I_{OS}§	V_{CC} = MAX		-40		-100	mA
I_{CC}	V_{CC} = MAX, Outputs enabled and open	SN54S'	60		74	mA
		SN74S'	75		90	

†For conditions shown as MIN or MAX, use the appropriate value specified under recommended operating conditions.
‡All typical values are at V_{CC} = 5 V, T_A = 25°C.
§Not more than one output should be shorted at a time, and duration of the short circuit test should not exceed one second.

switching characteristics, V_{CC} = 5 V, T_A = 25°C (see note 2)

PARAMETER¶	FROM (INPUT)	TO (OUTPUT)	LEVELS OF DELAY	TEST CONDITIONS	SN54S139 SN74S139 MIN	TYP	MAX	UNIT
t_{PLH}	Binary Select	Any	2			5	7.5	ns
t_{PHL}						6.5	10	ns
t_{PLH}			3	R_L = 280 Ω, C_L = 15 pF		7	12	ns
t_{PHL}						8	12	ns
t_{PLH}	Enable	Any	2			5	8	ns
t_{PHL}						6.5	10	ns

¶ t_{PLH} = propagation delay time, low to high-level output; t_{PHL} = propagation delay time, high to low-level output.
NOTE 2: See General Information Section for load circuits and voltage waveforms.

TEXAS INSTRUMENTS
POST OFFICE BOX 225012 ● DALLAS, TEXAS 75265

TYPES SN54LS139A, SN74LS139A
DUAL 2-LINE TO 4-LINE DECODERS/DEMULTIPLEXERS

recommended operating conditions

	SN54LS139A MIN	NOM	MAX	SN74LS139A MIN	NOM	MAX	UNIT
V_{CC} Supply voltage	4.5	5	5.5	4.75	5	5.25	V
V_{IH} High-level input voltage	2			2			V
V_{IL} Low-level input voltage			0.7			0.8	V
I_{OH} High-level output current			-0.4			-0.4	mA
I_{OL} Low-level output current			4			8	mA
T_A Operating free-air temperature	-55		125	0		70	°C

electrical characteristics over recommended operating free-air temperature range (unless otherwise noted)

PARAMETER	TEST CONDITIONS†		SN54LS139A SN74LS139A‡ MIN	TYP‡	MAX	UNIT
V_{IK}	V_{CC} = MIN, I_I = -18 mA				-1.5	V
V_{OH}	V_{CC} = MIN, V_{IH} = 2 V, V_{IL} = MAX, I_{OH} = -0.4 mA		2.5	3.4		V
V_{OL}	V_{CC} = MIN, V_{IH} = 2 V, V_{IL} = MAX	I_{OL} = 4 mA		0.25	0.4	V
		I_{OL} = 8 mA		0.35	0.5	
I_I	V_{CC} = MAX, V_I = 7 V				0.1	mA
I_{IH}	V_{CC} = MAX, V_I = 2.7 V				20	µA
I_{IL}	V_{CC} = MAX, V_I = 0.4 V				-0.4	mA
I_{OS}§	V_{CC} = MAX		-20		-100	mA
I_{CC}	V_{CC} = MAX, Outputs enabled and open		6.8		11	mA

†For conditions shown as MIN or MAX, use the appropriate value specified under recommended operating conditions.
‡All typical values are at V_{CC} = 5 V, T_A = 25°C.

switching characteristics, V_{CC} = 5 V, T_A = 25°C (see note 2)

PARAMETER¶	FROM (INPUT)	TO (OUTPUT)	LEVELS OF DELAY	TEST CONDITIONS	SN54LS139A SN74LS139A MIN	TYP	MAX	UNIT
t_{PLH}	Binary Select	Any	2			13	20	ns
t_{PHL}						22	33	ns
t_{PLH}			3	R_L = 2 kΩ, C_L = 15 pF		18	29	ns
t_{PHL}						25	38	ns
t_{PLH}	Enable	Any	2			16	24	ns
t_{PHL}						21	32	ns

¶ t_{PLH} = propagation delay time, low to high-level output; t_{PHL} = propagation delay time, high-to-low-level output.
NOTE 2: See General Information Section for load circuits and voltage waveforms.

TEXAS INSTRUMENTS
POST OFFICE BOX 225012 ● DALLAS, TEXAS 75265

TYPES SN54150, SN54151A, SN54152A, SN54LS151, SN54LS152, SN54S151, SN74150, SN74LS151, SN74S151 DATA SELECTORS/MULTIPLEXERS

DECEMBER 1972—REVISED DECEMBER 1983

- '150 Selects One-of-Sixteen Data Sources
- Others Select One-of-Eight Data Sources
- Performs Parallel-to-Serial Conversion
- Permits Multiplexing from N Lines to One Line
- Also For Use as Boolean Function Generator
- Input-Clamping Diodes Simplify System Design
- Fully Compatible with Most TTL Circuits

TYPE	TYPICAL AVERAGE PROPAGATION DELAY TIME DATA INPUT TO W OUTPUT	TYPICAL POWER DISSIPATION
'150	13 ns	200 mW
'151A	8 ns	145 mW
'152A	8 ns	130 mW
'LS151	13 ns	30 mW
'LS152	13 ns	28 mW
'S151	4.5 ns	225 mW

SN54150 . . . J OR W PACKAGE
SN74150 . . . J OR N PACKAGE
(TOP VIEW)

E7	1	24 V_CC
E6	2	23 E8
E5	3	22 E9
E4	4	21 E10
E3	5	20 E11
E2	6	19 E12
E1	7	18 E13
E0	8	17 E14
$\bar{G}$	9	16 E15
W	10	15 A
D	11	14 B
GND	12	13 C

SN54151A, SN54S151 . . . J OR W PACKAGE
SN74151A, SN74S151 . . . D, J OR N PACKAGE
(TOP VIEW)

D3	1	16 V_CC
D2	2	15 D4
D1	3	14 D5
D0	4	13 D6
Y	5	12 D7
W	6	11 A
$\bar{G}$	7	10 B
GND	8	9 C

SN54LS151, SN54S151 . . . FK PACKAGE
SN74LS151, SN74S151 . . . FN PACKAGE
(TOP VIEW)

NC — No internal connection

SN54152A, SN54LS152 . . . W PACKAGE
(TOP VIEW)

D4	1	14 V_CC
D3	2	13 D5
D2	3	12 D6
D1	4	11 D7
D0	5	10 A
W	6	9 B
GND	7	8 C

description

These monolithic data selectors/multiplexers contain full on-chip binary decoding to select the desired data source. The '150 selects one-of-sixteen data sources; the '151A, '152A, 'LS151, 'LS152, and 'S151 select one-of-eight data sources. The '150, '151A, 'LS151, and 'S151 have a strobe input which must be at a low logic level to enable these devices. A high level at the strobe forces the W output high, and the Y output (as applicable) low.

The '151A, 'LS151, and 'S151 feature complementary W and Y outputs whereas the '150, '152A, and 'LS152 have an inverted (W) output only.

The '151A and '152A incorporate address buffers which have symmetrical propagation delay times through the complementary paths. This reduces the possibility of transients occurring at the output(s) due to changes made at the select inputs, even when the '151A outputs are enabled (i.e., strobe low).

For SN54LS152 Chip Carrier Information, Contact The Factory.

schematics of inputs and outputs

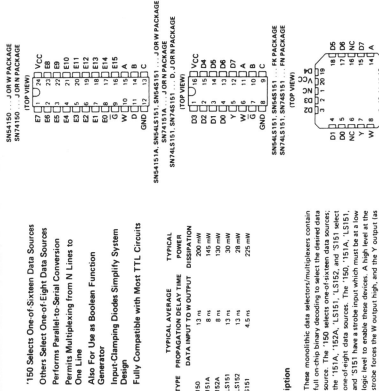

TYPICAL OF ALL OUTPUTS OF '150, '151A, '152A (130 Ω NOM, V_CC, OUTPUT)

TYPICAL OF ALL OUTPUTS OF 'LS151, 'LS152 (120 Ω NOM, V_CC, OUTPUT)

TYPICAL OF ALL OUTPUTS OF 'S151 (50 Ω NOM, V_CC, OUTPUT)

logic

'150 FUNCTION TABLE

SELECT INPUTS D	C	B	A	STROBE $\bar{G}$	OUTPUT W
X	X	X	X	H	H
L	L	L	L	L	$\bar{E}0$
L	L	L	H	L	$\bar{E}1$
L	L	H	L	L	$\bar{E}2$
L	L	H	H	L	$\bar{E}3$
L	H	L	L	L	$\bar{E}4$
L	H	L	H	L	$\bar{E}5$
L	H	H	L	L	$\bar{E}6$
L	H	H	H	L	$\bar{E}7$
H	L	L	L	L	$\bar{E}8$
H	L	L	H	L	$\bar{E}9$
H	L	H	L	L	$\bar{E}10$
H	L	H	H	L	$\bar{E}11$
H	H	L	L	L	$\bar{E}12$
H	H	L	H	L	$\bar{E}13$
H	H	H	L	L	$\bar{E}14$
H	H	H	H	L	$\bar{E}15$

'151A, 'LS151, 'S151 FUNCTION TABLE

SELECT INPUTS C	B	A	STROBE $\bar{G}$	OUTPUTS Y	W
X	X	X	H	L	H
L	L	L	L	D0	$\bar{D}0$
L	L	H	L	D1	$\bar{D}1$
L	H	L	L	D2	$\bar{D}2$
L	H	H	L	D3	$\bar{D}3$
H	L	L	L	D4	$\bar{D}4$
H	L	H	L	D5	$\bar{D}5$
H	H	L	L	D6	$\bar{D}6$
H	H	H	L	D7	$\bar{D}7$

'152A, 'LS152 FUNCTION TABLE

SELECT INPUTS C	B	A	OUTPUT W
L	L	L	$\bar{D}0$
L	L	H	$\bar{D}1$
L	H	L	$\bar{D}2$
L	H	H	$\bar{D}3$
H	L	L	$\bar{D}4$
H	L	H	$\bar{D}5$
H	H	L	$\bar{D}6$
H	H	H	$\bar{D}7$

H = high level, L = low level, X = irrelevant
E0, E1 . . . E15 = the complement of the level of the respective E input
D0, D1 . . . D7 = the level of the D respective input

PRODUCTION DATA
This document contains information current as of publication date. Products conform to specifications per the terms of Texas Instruments standard warranty. Production processing does not necessarily include testing of all parameters.

TEXAS INSTRUMENTS
POST OFFICE BOX 225012 • DALLAS, TEXAS 75265

SN54153, SN54LS153, SN54S153
SN74153, SN74LS153, SN74S153
DUAL 4-LINE TO 1-LINE DATA SELECTORS/MULTIPLEXERS

DECEMBER 1972 — REVISED MARCH 1988

- Permits Multiplexing from N lines to 1 line
- Performs Parallel-to-Serial Conversion
- Strobe (Enable) Line Provided for Cascading (N lines to n lines)
- High-Fan-Out, Low-Impedance, Totem-Pole Outputs
- Fully Compatible with most TTL Circuits

TYPE	TYPICAL AVERAGE PROPAGATION DELAY TIMES			TYPICAL POWER DISSIPATION
	FROM DATA	FROM STROBE	FROM SELECT	
'153	14 ns	17 ns	22 ns	180 mW
'LS153	14 ns	19 ns	22 ns	31 mW
'S153	6 ns	9.5 ns	12 ns	225 mW

description

Each of these monolithic, data selectors/multiplexers contains inverters and drivers to supply fully complementary, on-chip, binary decoding data selection to the AND-OR gates. Separate strobe inputs are provided for each of the two four-line sections.

SN54153, SN54LS153, SN54S153 ... J OR W PACKAGE
SN74153 ... N PACKAGE
SN74LS153, SN74S153 ... D OR N PACKAGE
(TOP VIEW)

```
    1G̅ [ 1      16 ] Vcc
     B [ 2      15 ] 2G̅
   1C3 [ 3      14 ] A
   1C2 [ 4      13 ] 2C3
   1C1 [ 5      12 ] 2C2
   1C0 [ 6      11 ] 2C1
    1Y [ 7      10 ] 2C0
   GND [ 8       9 ] 2Y
```

SN54LS153, SN54S153 ... FK PACKAGE
(TOP VIEW)

NC - No internal connection

FUNCTION TABLE

SELECT INPUTS		DATA INPUTS				STROBE	OUTPUT
B	A	C0	C1	C2	C3	G̅	Y
X	X	X	X	X	X	H	L
L	L	L	X	X	X	L	L
L	L	H	X	X	X	L	H
L	H	X	L	X	X	L	L
L	H	X	H	X	X	L	H
H	L	X	X	L	X	L	L
H	L	X	X	H	X	L	H
H	H	X	X	X	L	L	L
H	H	X	X	X	H	L	H

Select inputs A and B are common to both sections.
H = high level, L = low level, X = irrelevant

absolute maximum ratings over operating free-air temperature range (unless otherwise noted)

Supply voltage, Vcc (See Note 1) .... 7 V
Input voltage: '153, 'S153 .... 5.5 V
        'LS153 .... 7 V
Operating free-air temperature range: SN54' .... −55°C to 125°C
        SN74' .... 0°C to 70°C
Storage temperature range .... −65°C to 150°C

NOTE 1: Voltage values are with respect to network ground terminal.

TEXAS INSTRUMENTS
POST OFFICE BOX 655012 • DALLAS, TEXAS 75265

SN54153, SN54LS153, SN54S153
SN74153, SN74LS153, SN74S153
DUAL 4-LINE TO 1-LINE DATA SELECTORS/MULTIPLEXERS

logic symbol†

```
A  (14)
B  (2)

        0 G 3
        1

1G̅  (1)      EN
1C0 (6)       MUX
1C1 (5)       0
1C2 (4)       1
1C3 (3)       2
              3
2G̅ (15)
2C0 (10)
2C1 (11)
2C2 (12)
2C3 (13)
                    (7) 1Y
                    (9) 2Y
```

†This symbol is in accordance with ANSI/IEEE Std. 91-1984 and IEC Publication 617-12.

logic diagrams (positive logic)

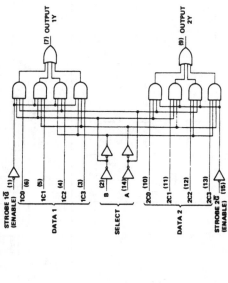

Pin numbers shown are for D, J, N, and W packages.

TEXAS INSTRUMENTS
POST OFFICE BOX 225012 • DALLAS, TEXAS 75265

TYPES SN54174, SN54175, SN54LS174, SN54LS175, SN54S174, SN54S175, SN74174, SN74175, SN74LS174, SN74LS175, SN74S174, SN74S175 HEX/QUADRUPLE D-TYPE FLIP-FLOPS WITH CLEAR

DECEMBER 1972–REVISED DECEMBER 1983

- '174, 'LS174, 'S174 . . . HEX D-TYPE FLIP-FLOPS
- '175, 'LS175, 'S175 . . . QUADRUPLE D-TYPE FLIP-FLOPS

- **'174, 'LS174, 'S174 Contain Six Flip-Flops with Single-Rail Outputs**
- **'175, 'LS175, 'S175 Contain Four Flip-Flops with Double-Rail Outputs**
- **Three Performance Ranges Offered: See Table Lower Right**
- **Buffered Clock and Direct Clear Inputs**
- **Individual Data Input to Each Flip-Flop**
- **Applications include:**
  Buffer/Storage Registers
  Shift Registers
  Pattern Generators

description

These monolithic, positive-edge-triggered flip-flops utilize TTL circuitry to implement D-type flip-flop logic. All have a direct clear input, and the '175, 'LS175, and 'S175 feature complementary outputs from each flip-flop.

Information at the D inputs meeting the setup time requirements is transferred to the Q outputs on the positive-going edge of the clock pulse. Clock triggering occurs at a particular voltage level and is not directly related to the transition time of the positive-going pulse. When the clock input is at either the high or low level, the D input signal has no effect at the output.

These circuits are fully compatible for use with most TTL circuits.

FUNCTION TABLE
(EACH FLIP-FLOP)

INPUTS			OUTPUTS	
CLEAR	CLOCK	D	Q	$\bar{Q}$†
L	X	X	L	H
H	↑	H	H	L
H	↑	L	L	H
H	L	X	Q_0	$\bar{Q}_0$

H = high level (steady state)
L = low level (steady state)
X = irrelevant
↑ = transition from low to high level
Q_0 = the level of Q before the indicated steady-state input conditions were established.
† = '175, 'LS175, and 'S175 only.

TYPES	TYPICAL MAXIMUM CLOCK FREQUENCY	TYPICAL POWER DISSIPATION PER FLIP-FLOP
'174, '175	35 MHz	38 mW
'LS174, 'LS175	40 MHz	14 mW
'S174, 'S175	110 MHz	75 mW

SN54174, SN54LS174, SN54S174 . . . J OR W PACKAGE
SN74174 . . . J OR N PACKAGE
SN74LS174, SN74S174 . . . D, J OR N PACKAGE
(TOP VIEW)

CLR	1	16	VCC
1Q	2	15	6Q
1D	3	14	6D
2D	4	13	5D
2Q	5	12	5Q
3D	6	11	4D
3Q	7	10	4Q
GND	8	9	CLK

SN54LS174, SN54S174 . . . FK PACKAGE
SN54LS174, SN54S174 . . . FN PACKAGE
(TOP VIEW)

SN54175, SN54LS175, SN54S175 . . . J OR W PACKAGE
SN74175 . . . J OR N PACKAGE
SN74LS175, SN74S175 . . . D, J OR N PACKAGE
(TOP VIEW)

CLR	1	16	VCC
1Q	2	15	4Q
1Q̄	3	14	4Q̄
1D	4	13	4D
2D	5	12	3D
2Q̄	6	11	3Q̄
2Q	7	10	3Q
GND	8	9	CLK

SN54LS175, SN54S175 . . . FK PACKAGE
SN54LS175, SN54S175 . . . FN PACKAGE
(TOP VIEW)

NC – No internal connection

logic diagrams

'174, 'LS174, 'S174

'175, 'LS175, 'S175

Pin numbers shown on logic notation are for D, J or N packages.

TEXAS INSTRUMENTS
POST OFFICE BOX 225012 • DALLAS, TEXAS 75265

absolute maximum ratings (Note 1)

Supply Voltage, V_{CC}	7V
Input Voltage	5.5V
Output Voltage	5.5V
Storage Temperature Range	−65°C to +150°C
Lead Temperature (Soldering, 10 seconds)	300°C

operating conditions

		MIN	MAX	UNITS
Supply Voltage (V_{CC})	DM54LS189	4.5	5.5	V
	DM74LS189	4.75	5.25	V
Temperature (T_A)	DM54LS189	−55	+125	°C
	DM74LS189	0	+70	°C

electrical characteristics

Over recommended operating free-air temperature range (unless otherwise noted) (Notes 2 and 3)

	PARAMETER	CONDITIONS		MIN	TYP	MAX	UNITS
V_{IH}	High Level Input Voltage			2			V
V_{IL}	Low Level Input Voltage					0.8	V
V_{OH}	High Level Output Voltage	V_{CC} = Min, I_{OH} = −2 mA		2.4	3.4		V
				2.4	3.2		
V_{OL}	Low Level Output Voltage	V_{CC} = Min	I_{OL} = 4 mA DM54LS189			0.45	V
			I_{OL} = 8 mA DM74LS189			0.5	
I_{IH}	High Level Input Current	V_{CC} = Max, V_I = 2.7				10	µA
I_I	High Level Input Current at Maximum Voltage	V_{CC} = Max, V_I = 5.5V				1.0	mA
I_{IL}	Low Level Input Current	V_{CC} = Max, V_I = 0.45V				−100	µA
I_{OS}	Short-Circuit Output Current (Note 4)	V_{CC} = Max, V_O = 0V		−30		−100	mA
I_{CC}	Supply Current (Note 5)	V_{CC} = Max			15	29	mA
V_{IC}	Input Clamp Voltage	V_{CC} = Min, I_I = −18 mA				−1.2	V
I_{OZH}	TRI-STATE Output Current, High Level Voltage Applied	V_{CC} = Max, V_O = 2.4V				40	µA
I_{OZL}	TRI-STATE Output Current, Low Level Voltage Applied	V_{CC} = Max, V_O = 0.45V				40	µA

switching time waveforms

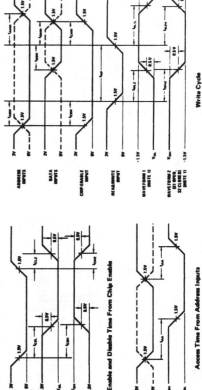

Access Time From Address Inputs

Enable and Disable Time From Chip Enable

Write Cycle

FIGURE 1

Note 1: Waveform 1 is for the output with internal conditions such that the output is low except when disabled. Waveform 2 is for the output with internal conditions such that the output is high except when disabled.

Note 2: When measuring delay times from address inputs, the chip enable input is low and the read/write input is high.

Note 3: When measuring delay times from chip enable input, the address inputs are steady-state and the read/write input is high.

Note 4: Input waveforms are supplied by pulse generators having the following characteristics: $t_r \leq 2.5$ ns, $t_f \leq 2.5$ ns, PRR ≤ 1 MHz, and $Z_{OUT} = 50\Omega$.

DM54LS189/DM74LS189 low power 64-bit random access memories with TRI-STATE® outputs

general description

These 64-bit active-element memories are monolithic Schottky-clamped transistor-transistor logic (TTL) arrays organized as 16 words of 4 bits each. They are fully decoded and feature a chip enable input to simplify decoding required to achieve the desired system organization. This device is implemented with low power Schottky technology resulting in one-fifth power while retaining the speed of standard TTL.

The TRI-STATE output combines the convenience of an open-collector with the speed of a totem-pole output; it can be bus-connected to other similar outputs, yet it retains the fast rise time characteristics of the TTL totem-pole output. Systems utilizing data bus lines with a defined pull-up impedance can employ the open-collector DM54LS289.

but it will allow the bus line to be driven by another active output or a passive pull-up if desired.

Read Cycle: The stored information (complement of information applied at the data inputs during the write cycle) is available at the outputs when the read/write input is high and the chip enable is low. When the chip enable input is high, the outputs will be in the high impedance state.

Write Cycle: The complement of the information at the data input is written into the selected location when both the chip enable input and the read/write input are low. While the read/write input is low, the outputs are in the high impedance state. When a number of the DM54LS189 outputs are bus-connected, this high impedance state will neither load nor drive the bus line,

features

- Schottky-clamped for high speed applications
  - Access from chip enable input—40 ns typ
  - Access from address inputs—60 ns typ
- TRI-STATE outputs drive bus-organized systems and/or high capacitive loads
- Low power—75 mW typ
- DM54LS189 is guaranteed for operation over the full military temperature range of −55°C to +125°C
- Compatible with most TTL and DTL logic circuits
- Chip enable input simplifies system decoding

truth table

FUNCTION	INPUTS		OUTPUT
	CHIP ENABLE	READ/WRITE	
Write (Store Complement of Data)	L	L	Hz
Read	L	H	Stored Data
Inhibit	H	X	Hz

H = high level
L = low level
X = don't care

Order Number DM54LS189J or DM74LS189J
See Package 10
Order Number DM74LS189N
See Package 15
Order Number DM54LS189W
See Package 28

connection diagram

Dual-In-Line and Flat Package

TOP VIEW

TYPES SN54190, SN54191, SN54LS190, SN54LS191,
SN74190, SN74191, SN74LS190, SN74LS191
SYNCHRONOUS UP/DOWN COUNTERS WITH DOWN/UP MODE CONTROL

DECEMBER 1972—REVISED DECEMBER 1983

- Counts 8-4-2-1 BCD or Binary
- Single Down/Up Count Control Line
- Count Enable Control Input
- Ripple Clock Output for Cascading
- Asynchronously Presettable with Load Control
- Parallel Outputs
- Cascadable for n-Bit Applications

TYPE	AVERAGE PROPAGATION DELAY	TYPICAL MAXIMUM CLOCK FREQUENCY	TYPICAL POWER DISSIPATION
'190,'191	20ns	25MHz	325mW
'LS190,'LS191	20ns	25MHz	100mW

description

The '190, 'LS190, '191, and 'LS191 are synchronous, reversible up/down counters having a complexity of 58 equivalent gates. The '191 and 'LS191 are 4-bit binary counters and the '190 and 'LS190 are BCD counters. Synchronous operation is provided by having all flip-flops clocked simultaneously so that the outputs change coincident with each other when so instructed by the steering logic. This mode of operation eliminates the output counting spikes normally associated with asynchronous (ripple clock) counters.

The outputs of the four master-slave flip-flops are triggered on a low-to-high transition of the clock input if the enable input is low. A high at the enable input inhibits counting. Level changes at the enable input should be made only when the clock input is high. The direction of the count is determined by the level of the down/up input. When low, the counter count up and when high, it counts down. A false clock may occur if the down/up input changes while the clock is low. A false ripple carry may occur if both the clock and enable are low and the down/up input is high during a load pulse.

These counters are fully programmable; that is, the outputs may be preset to either level by placing a low on the load input and entering the desired data at the data inputs. The output will change to agree with the data inputs independently of the level of the clock input. This feature allows the counters to be used as modulo-N dividers by simply modifying the count length with the preset inputs.

The clock, down/up, and load inputs are buffered to lower the drive requirement which significantly reduces the number of clock drivers, etc., required for long parallel words.

Two outputs have been made available to perform the cascading function: ripple clock and maximum/minimum count. The latter output produces a high-level output pulse with a duration approximately equal to one complete cycle of the clock when the counter overflows or underflows. The ripple clock output produces a low-level output pulse equal in width to the low-level portion of the clock input when an overflow or underflow condition exists. The counters can be easily cascaded by feeding the ripple clock output to the enable input of the succeeding counter if parallel clocking is used, or to the clock input if parallel enabling is used. The maximum/minimum count output can be used to accomplish look-ahead for high-speed operation.

Series 54' and 54LS' are characterized for operation over the full military temperature range of   55°C to 125°C; Series 74' and 74LS' are characterized for operation from 0°C to 70°C.

SN54190, SN54191, SN54LS190,
SN54LS191 . . . J OR W PACKAGE
SN74190, SN74191 . . . J OR N PACKAGE
SN74LS190, SN74LS191 . . . D, J OR N PACKAGE
(TOP VIEW)

B	1 16	VCC
QB	2 15	A
QA	3 14	CLK
CTEN	4 13	RCO
D/U	5 12	MAX/MIN
QC	6 11	LOAD
QD	7 10	C
GND	8 9	D

SN54LS190, SN54LS191 . . . FK PACKAGE
SN74LS190, SN74LS191 . . . FN PACKAGE
(TOP VIEW)

NC - No internal connection

TEXAS
INSTRUMENTS
POST OFFICE BOX 225012 ● DALLAS, TEXAS 75265

logic diagram

'191, 'LS191 BINARY COUNTERS

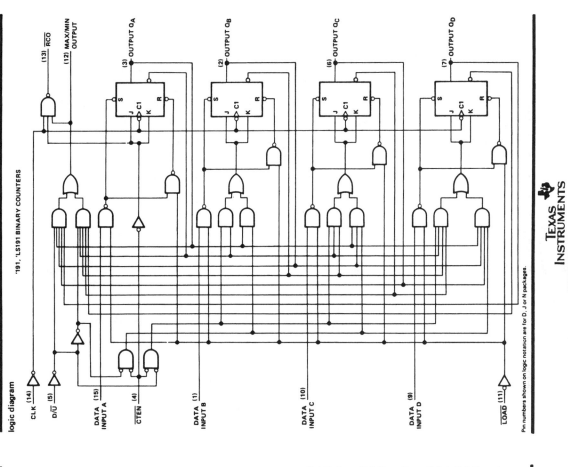

Pin numbers shown on logic notation are for D, J or N packages.

TEXAS
INSTRUMENTS
POST OFFICE BOX 225012 ● DALLAS, TEXAS 75265

TYPES SN54195, SN54LS195A, SN54S195, SN74195, SN74LS195A, SN74S195 4-BIT PARALLEL-ACCESS SHIFT REGISTERS

logic diagram

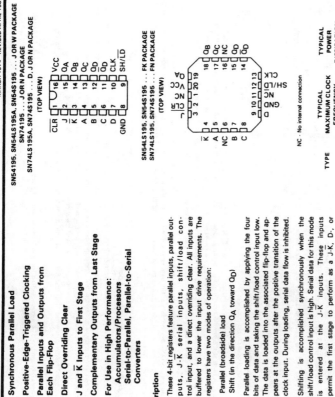

†This connection is made on '195 only.
Pin numbers shown on logic notation are for D, J or N packages.

TEXAS INSTRUMENTS
POST OFFICE BOX 225012 • DALLAS, TEXAS 75265

MARCH 1974 — REVISED APRIL 1985

- Synchronous Parallel Load
- Positive-Edge-Triggered Clocking
- Parallel Inputs and Outputs from Each Flip-Flop
- Direct Overriding Clear
- J and $\overline{K}$ Inputs to First Stage
- Complementary Outputs from Last Stage
- For Use in High Performance:
  Accumulators/Processors
  Serial-to-Parallel, Parallel-to-Serial
  Converters

description

These 4-bit registers feature parallel inputs, parallel outputs, J-$\overline{K}$ serial inputs, shift/load control input, and a direct overriding clear. All inputs are buffered to lower the input drive requirements. The registers have two modes of operation:

Parallel (broadside) load

Shift (in the direction Q_A toward Q_D)

Parallel loading is accomplished by applying the four bits of data and taking the shift/load control input low. The data is loaded into the associated flip-flop and appears at the outputs after the positive transition of the clock input. During loading, serial data flow is inhibited.

Shifting is accomplished synchronously when the shift/load control input is high. Serial data for this mode is entered at the J-$\overline{K}$ inputs. These inputs permit the first stage to perform as a J-K, D, or T-type flip-flop as shown in the function table.

The high-performance 'S195, with a 105-megahertz typical maximum shift-frequency, is particularly attractive for very-high-speed data processing systems. In most cases existing systems can be upgraded merely by using this Schottky-clamped shift register.

SN54195, SN54LS195A, SN54S195 . . . J OR W PACKAGE
SN74195 . . . J OR N PACKAGE
SN74LS195A, SN74S195 . . . D, J OR N PACKAGE
(TOP VIEW)

$\overline{CLR}$	1	16	V_{CC}
J	2	15	Q_A
$\overline{K}$	3	14	Q_B
A	4	13	Q_C
B	5	12	Q_D
C	6	11	$\overline{Q}_D$
D	7	10	CLK
GND	8	9	SH/$\overline{LD}$

SN54LS195, SN54S195 . . . FK PACKAGE
SN74LS195, SN74S195 . . . FN PACKAGE
(TOP VIEW)

NC — No internal connection

TYPE	TYPICAL MAXIMUM CLOCK FREQUENCY	TYPICAL POWER DISSIPATION
'195	39 MHz	195 mW
'LS195A	39 MHz	70 mW
'S195	105 MHz	350 mW

FUNCTION TABLE

INPUTS							OUTPUTS				
CLEAR	SHIFT/LOAD	CLOCK	SERIAL		PARALLEL				Q_A	Q_B	Q_C	Q_D	$\overline{Q}_D$

CLEAR	SHIFT/ LOAD	CLOCK	J	$\overline{K}$	A	B	C	D	Q_A	Q_B	Q_C	Q_D	$\overline{Q}_D$
L	X	X	X	X	X	X	X	X	L	L	L	L	H
H	L	↑	X	X	a	b	c	d	a	b	c	d	$\overline{d}$
H	H	L	X	X	X	X	X	X	Q_{A0}	Q_{B0}	Q_{C0}	Q_{D0}	$\overline{Q}_{D0}$
H	H	↑	L	H	X	X	X	X	Q_{A0}	Q_{An}	Q_{Bn}	Q_{Cn}	$\overline{Q}_{Cn}$
H	H	↑	L	L	X	X	X	X	L	Q_{An}	Q_{Bn}	Q_{Cn}	$\overline{Q}_{Cn}$
H	H	↑	H	H	X	X	X	X	H	Q_{An}	Q_{Bn}	Q_{Cn}	$\overline{Q}_{Cn}$
H	H	↑	H	L	X	X	X	X	$\overline{Q}_{An}$	Q_{An}	Q_{Bn}	Q_{Cn}	$\overline{Q}_{Cn}$

H = high level (steady state)
L = low level (steady state)
X = irrelevant (any input, including transitions)
↑ = transition from low to high level
a, b, c, d = the level of steady state input at A, B, C, or D, respectively.
Q_{A0}, Q_{B0}, Q_{C0}, Q_{D0} = the level of Q_A, Q_B, Q_C, or Q_D, respectively, before the indicated steady state input conditions were established.
Q_{An}, Q_{Bn}, Q_{Cn} = the level of Q_A, Q_B, or Q_C, respectively, before the most recent transition of the clock

TEXAS INSTRUMENTS
POST OFFICE BOX 225012 • DALLAS, TEXAS 75265

Courtesy of Texas Instruments Incorporated

National Semiconductor

CD4069M/CD4069C Inverter Circuits

General Description

The CD4069B consists of six inverter circuits and is manufactured using complementary MOS (CMOS) to achieve wide power supply operating range, low power consumption, high noise immunity, and symmetric controlled rise and fall times.

This device is intended for all general purpose inverter applications where the special characteristics of the MM74C901, MM74C903, MM74C907, and CD4049A Hex Inverter/Buffers are not required. In those applications requiring larger noise immunity the MM74C14 or MM74C914 Hex Schmitt Trigger is suggested.

All inputs are protected from damage due to static discharge by diode clamps to V_{DD} and V_{SS}.

Features

- Wide supply voltage range: 3.0V to 15V
- High noise immunity: $0.45 V_{DD}$ typ.
- Low power TTL compatibility: fan out of 2 driving 74L or 1 driving 74LS
- Equivalent to MM54C04/MM74C04

Absolute Maximum Ratings

(Notes 1 and 2)

V_{DD} dc Supply Voltage	−0.5 to +18 V_{DC}
V_{IN} Input Voltage	−0.5 to V_{DD} +0.5 V_{DC}
T_S Storage Temperature Range	−65°C to +150°C
P_D Package Dissipation	500 mW
T_L Lead Temperature (Soldering, 10 seconds)	300°C

Recommended Operating Conditions

(Note 2)

V_{DD} dc Supply Voltage	3 to 15 V_{DC}
V_{IN} Input Voltage	0 to V_{DD} V_{DC}
T_A Operating Temperature Range	
CD4069M	−55°C to +126°C
CD4069C	−40°C to +85°C

DC Electrical Characteristics

CD4069M (Note 2)

PARAMETER	CONDITIONS	−55°C MIN	−55°C MAX	25°C MIN	25°C TYP	25°C MAX	125°C MIN	125°C MAX	UNITS
I_{DD} Quiescent Device Current	V_{DD} = 5V		0.25			0.25		7.5	μA
	V_{DD} = 10V		0.5			0.5		15	μA
	V_{DD} = 15V		1.0			1.0		30	μA
V_{OL} Low Level Output Voltage	$I_{IO} < 1\mu A$								
	V_{DD} = 5V		0.05		0	0.05		0.05	V
	V_{DD} = 10V		0.05		0	0.05		0.05	V
	V_{DD} = 15V		0.05		0	0.05		0.05	V
V_{OH} High Level Output Voltage	$I_{IO} < 1\mu A$								
	V_{DD} = 5V	4.95		4.95			4.95		V
	V_{DD} = 10V	9.95		9.95			9.95		V
	V_{DD} = 15V	14.95		14.95			14.95		V
V_{IL} Low Level Input Voltage	$I_{IO} < 1\mu A$								
	V_{DD} = 5V, V_O = 4.5V		1.5			1.5		1.5	V
	V_{DD} = 10V, V_O = 9V		3.0			3.0		3.0	V
	V_{DD} = 15V, V_O = 13.5V		4.0			4.0		4.0	V
V_{IH} High Level Input Voltage	$I_{IO} < 1\mu A$								
	V_{DD} = 5V, V_O = 0.5V	3.5		3.5			3.5		V
	V_{DD} = 10V, V_O = 1V	7.0		7.0			7.0		V
	V_{DD} = 15V, V_O = 1.5V	11.0		11.0			11.0		V
I_{OL} Low Level Output Current	V_{DD} = 5V, V_O = 0.4V	0.64		0.51	0.88		0.36		mA
	V_{DD} = 10V, V_O = 0.5V	1.6		1.3	2.25		0.9		mA
	V_{DD} = 15V, V_O = 1.5V	4.2		3.4	8.8		2.4		mA
I_{OH} High Level Output Current	V_{DD} = 5V, V_O = 4.6V	−0.64		−0.51	−0.88		−0.36		mA
	V_{DD} = 10V, V_O = 9.5V	−1.6		−1.3	−2.25		−0.9		mA
	V_{DD} = 15V, V_O = 13.5V	−4.2		−3.4	−8.8		−2.4		mA
I_{IN} Input Current	V_{DD} = 15V, V_{IN} = 0V	−0.10			-10^{-5}	−0.10		−1.0	μA
	V_{DD} = 15V, V_{IN} = 15V	0.10			10^{-5}	0.10		1.0	μA

Schematic and Connection Diagrams

Dual-In-Line Package

TOP VIEW

AC Test Circuits and Switching Time Waveforms

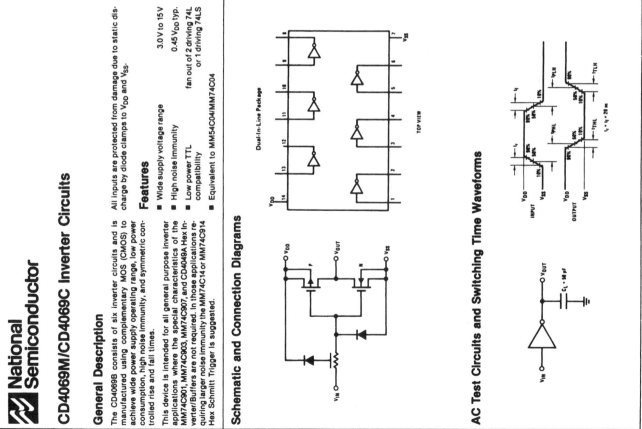

$t_r = t_f = 20$ ns

$C_L = 50$ pF

National Semiconductor

CD4071BM/CD4071BC Quad 2-Input OR Buffered B Series Gate
CD4081BM/CD4081BC Quad 2-Input AND Buffered B Series Gate

General Description

These quad gates are monolithic complementary MOS (CMOS) integrated circuits constructed with N- and P-channel enhancement mode transistors. They have equal source and sink current capabilities and conform to standard B series output drive. The devices also have buffered outputs which improve transfer characteristics by providing very high gain.

All inputs are protected against static discharge with diodes to V_{DD} and V_{SS}.

Features

■ Low power TTL compatibility   fan out of 2 driving 74L or 1 driving 74LS
■ 5V-10V-15V parametric ratings
■ Symmetrical output characteristics
■ Maximum input leakage 1μA at 15V over full temperature range

Absolute Maximum Ratings
(Notes 1 and 2)

Voltage at Any Pin	$-0.5V$ to $V_{DD} + 0.5V$
Package Dissipation	500 mW
V_{DD} Range	$-0.5\ V_{DC}$ to $+18\ V_{DC}$
Storage Temperature	$-65°C$ to $+150°C$
Lead Temperature (Soldering, 10 seconds)	260°C

Operating Conditions

Operating V_{DD} Range	3 V_{DC} to 15 V_{DC}
Operating Temperature Range	
CD4071BM, CD4081BM	$-55°C$ to $+125°C$
CD4071BC, CD4081BC	$-40°C$ to $+85°C$

DC Electrical Characteristics — CD4071BM/CD4081BM (Note 2)

SYM	PARAMETER	CONDITIONS	−55°C MIN	−55°C MAX	+25°C MIN	+25°C TYP	+25°C MAX	+125°C MIN	+125°C MAX	UNITS
I_{DD}	Quiescent Device Current	V_{DD} = 5V		0.25		0.004	0.25		7.5	μA
		V_{DD} = 10V		0.50		0.005	0.50		15	μA
		V_{DD} = 15V		1.0		0.006	1.0		30	μA
V_{OL}	Low Level Output Voltage	V_{DD} = 5V $\}$ $|I_{OL}| < 1\mu A$		0.05		0	0.05		0.05	V
		V_{DD} = 10V		0.05		0	0.05		0.05	V
		V_{DD} = 15V		0.05		0	0.05		0.05	V
V_{OH}	High Level Output Voltage	V_{DD} = 5V $\}$ $|I_{OH}| < 1\mu A$	4.95		4.95	5		4.95		V
		V_{DD} = 10V	9.95		9.95	10		9.95		V
		V_{DD} = 15V	14.95		14.95	15		14.95		V
V_{IL}	Low Level Input Voltage	V_{DD} = 5V, V_O = 0.5V		1.5		2	1.5		1.5	V
		V_{DD} = 10V, V_O = 1.0V		3.0		4	3.0		3.0	V
		V_{DD} = 15V, V_O = 1.5V		4.0		6	4.0		4.0	V
V_{IH}	High Level Input Voltage	V_{DD} = 5V, V_O = 4.5V	3.5		3.5	3		3.5		V
		V_{DD} = 10V, V_O = 9.0V	7.0		7.0	6		7.0		V
		V_{DD} = 15V, V_O = 13.5V	11.0		11.0	9		11.0		V
I_{OL}	Low Level Output Current (Note 3)	V_{DD} = 5V, V_O = 0.4V	0.64		0.51	0.88		0.36		mA
		V_{DD} = 10V, V_O = 0.5V	1.6		1.3	2.25		0.9		mA
		V_{DD} = 15V, V_O = 1.5V	4.2		3.4	8.8		2.4		mA
I_{OH}	High Level Output Current (Note 3)	V_{DD} = 5V, V_O = 4.6V	−0.64		−0.51	−0.88		−0.36		mA
		V_{DD} = 10V, V_O = 9.5V	−1.6		−1.3	−2.25		−0.9		mA
		V_{DD} = 15V, V_O = 13.5V	−4.2		−3.4	−8.8		−2.4		mA
I_{IN}	Input Current	V_{DD} = 15V, V_{IN} = 0V		−0.10		-10^{-5}	−0.10		−1.0	μA
		V_{DD} = 15V, V_{IN} = 15V		0.10		10^{-5}	0.10		1.0	μA

Note 1: "Absolute Maximum Ratings" are those values beyond which the safety of the device cannot be guaranteed. Except for "Operating Temperature Range" they are not meant to imply that the devices should be operated at these limits. The table of "Electrical Characteristics" provides conditions for actual device operation.

Note 2: All voltages measured with respect to V_{SS} unless otherwise specified.

Note 3: I_{OH} and I_{OL} are tested one output at a time.

Schematic and Connection Diagrams

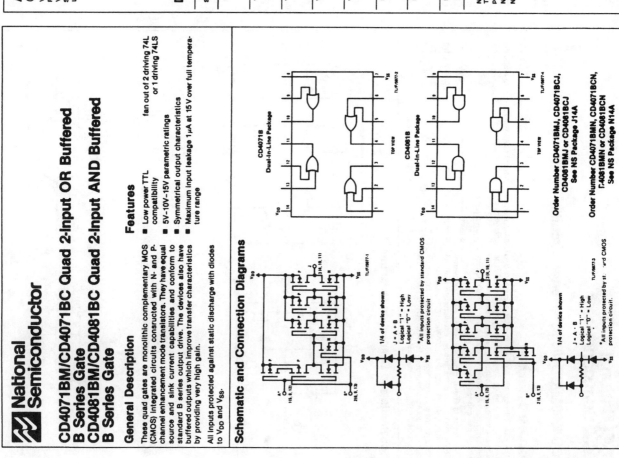

CD4071B Dual-In-Line Package
TOP VIEW

CD4081B Dual-In-Line Package
TOP VIEW

1/4 of device shown
J = A + B
Logical "1" = High
Logical "0" = Low
*All inputs protected by standard CMOS protection circuit.

1/4 of device shown
J = A · B
Logical "1" = High
Logical "0" = Low
*All inputs protected by standard CMOS protection circuit.

Order Number CD4071BMJ, CD4071BCJ, CD4081BMJ or CD4081BCJ
See NS Package J14A

Order Number CD4071BMN, CD4071BCN, CD4081BMN or CD4081BCN
See NS Package N14A

AC Electrical Characteristics CD4081BC/CD4081BM

$T_A = 25°C$, Input t_r; $t_f = 20$ ns, $C_L = 50$ pF. $R_L = 200K$ Typical temperature coefficient is 0.3%/°C

SYMBOL	PARAMETER	CONDITIONS	TYP	MAX	UNITS
tPHL	Propagation Delay Time, High-to-Low Level	VDD = 5V	100	250	ns
		VDD = 10V	40	100	ns
		VDD = 15V	30	70	ns
tPLH	Propagation Delay Time, Low-to-High Level	VDD = 5V	120	250	ns
		VDD = 10V	50	100	ns
		VDD = 15V	35	70	ns
tTHL, tTLH	Transition Time	VDD = 5V	90	200	ns
		VDD = 10V	50	100	ns
		VDD = 15V	40	80	ns
CIN	Average Input Capacitance	Any Input	5	7.5	pF
CPD	Power Dissipation Capacity	Any Gate	18		pF

Typical Performance Characteristics

FIGURE 1. Typical Transfer Characteristics

FIGURE 2. Typical Transfer Characteristics

FIGURE 3. Typical Transfer Characteristics

FIGURE 4. Typical Transfer Characteristics

FIGURE 5

FIGURE 6

DC Electrical Characteristics CD4071BC/CD4081BC (Note 2)

SYM	PARAMETER	CONDITIONS	-40°C MIN	-40°C MAX	+25°C MIN	+25°C TYP	+25°C MAX	+85°C MIN	+85°C MAX	UNITS
IDD	Quiescent Device Current	VDD = 5V		1		0.004	1		7.5	µA
		VDD = 10V		2		0.005	2		15	µA
		VDD = 15V		4		0.006	4		30	µA
VOL	Low Level Output Voltage	VDD = 5V		0.05		0	0.05		0.05	V
		VDD = 10V |IO| < 1µA		0.05		0	0.05		0.05	V
		VDD = 15V		0.05		0	0.05		0.05	V
VOH	High Level Output Voltage	VDD = 5V	4.95		4.95	5		4.95		V
		VDD = 10V |IO| < 1µA	9.95		9.95	10		9.95		V
		VDD = 15V	14.95		14.95	15		14.95		V
VIL	Low Level Input Voltage	VDD = 5V, VO = 0.5V		1.5		2	1.5		1.5	V
		VDD = 10V, VO = 1.0V		3.0		4	3.0		3.0	V
		VDD = 15V, VO = 1.5V		4.0		6	4.0		4.0	V
VIH	High Level Input Voltage	VDD = 5V, VO = 4.5V	3.5		3.5	3		3.5		V
		VDD = 10V, VO = 9.0V	7.0		7.0	6		7.0		V
		VDD = 15V, VO = 13.5V	11.0		11.0	9		11.0		V
IOL	Low Level Output Current (Note 3)	VDD = 5V, VO = 0.4V	0.52		0.44	0.88		0.36		mA
		VDD = 10V, VO = 0.5V	1.3		1.1	2.25		0.9		mA
		VDD = 15V, VO = 1.5V	3.6		3.0	8.8		2.4		mA
IOH	High Level Output Current (Note 3)	VDD = 5V, VO = 4.6V	-0.52		-0.44	-0.88		-0.36		mA
		VDD = 10V, VO = 9.5V	-1.3		-1.1	-2.25		-0.9		mA
		VDD = 15V, VO = 13.5V	-3.6		-3.0	-8.8		-2.4		mA
IIN	Input Current	VDD = 15V, VIN = 0V		-0.30		-10^{-5}	-0.30		-1.0	µA
		VDD = 15V, VIN = 15V		0.30		10^{-5}	0.30		1.0	µA

AC Electrical Characteristics CD4071BC/CD4071BM

$T_A = 25°C$, Input t_r; $t_f = 20$ ns, $C_L = 50$ pF. $R_L = 200K\Omega$ Typical temperature coefficient is 0.3%/°C

SYMBOL	PARAMETER	CONDITIONS	TYP	MAX	UNITS
tPHL	Propagation Delay Time, High-to-Low Level	VDD = 5V	100	250	ns
		VDD = 10V	40	100	ns
		VDD = 15V	30	70	ns
tPLH	Propagation Delay Time, Low-to-High Level	VDD = 5V	90	250	ns
		VDD = 10V	40	100	ns
		VDD = 15V	30	70	ns
tTHL, tTLH	Transition Time	VDD = 5V	90	200	ns
		VDD = 10V	50	100	ns
		VDD = 15V	40	80	ns
CIN	Average Input Capacitance	Any Input	5	7.5	pF
CPD	Power Dissipation Capacity	Any Gate	18		pF

Note 1: "Absolute Maximum Ratings" are those values beyond which the safety of the device cannot be guaranteed. Except for "Operating Temperature Range" they are not meant to imply that the devices should be operated at these limits. The table of "Electrical Characteristics" provides conditions for actual device operation.

Note 2: All voltages measured with respect to VSS unless otherwise specified.

Note 3: IOH and IOL are tested one output at a time.

MOTOROLA

MC14051B
MC14052B
MC14053B

CMOS MSI
(LOW-POWER COMPLEMENTARY MOS)

ANALOG MULTIPLEXERS/DEMULTIPLEXERS

CASE 620
L SUFFIX
CERAMIC PACKAGE

CASE 648
P SUFFIX
PLASTIC PACKAGE

ORDERING INFORMATION

A Series: -55°C to +125°C
MC14XXXBAL (Ceramic Package Only)

C Series: -40°C to +85°C
MC14XXXBCP (Plastic Package)
MC14XXXBCL (Ceramic Package)

ANALOG MULTIPLEXERS/DEMULTIPLEXERS

The MC14051B, MC14052B, and MC14053B analog multiplexers are digitally-controlled analog switches. The MC14051B effectively implements an SP8T solid state switch, the MC14052B a DP4T, and the MC14053B a Triple SPDT. All three devices feature low ON impedance and very low OFF leakage current. Control of analog signals up to the complete supply voltage range can be achieved.

- Diode Protection on All Inputs
- Supply Voltage Range = 3.0 Vdc to 18 Vdc
- Analog Voltage Range (VDD – VEE) = 3 to 18 V
  Note: VEE must be ≤ VSS
- Linearized Transfer Characteristics
- Low-Noise — 12 nV/√Cycle, f > 1 kHz typical
- Pin-for-Pin Replacement for CD4051, CD4052, and CD4053
- For 4PDT Switch, See MC14551B
- For Lower RON, Use the HC4051, HC4052, or HC4053 High-Speed CMOS Devices

MAXIMUM RATINGS*

Symbol	Parameter	Value	Unit
V_{DD}	DC Supply Voltage (Referenced to V_{EE}: $V_{SS} > V_{EE}$)	-0.5 to +18.0	V
V_{in}, V_{out}	Input or Output Voltage (DC or Transient) (Referenced to V_{SS} for Control Inputs and V_{EE} for Switch I/O)	-0.5 to V_{DD} + 0.5	V
I_{in}	Input Current (DC or Transient), per Control Pin	±10	mA
I_{sw}	Switch Through Current	±25	mA
P_D	Power Dissipation, per Package†	500	mW
T_{stg}	Storage Temperature	-65 to +150	°C
T_L	Lead Temperature (8-Second Soldering)	260	°C

*Maximum Ratings are those values beyond which damage to the device may occur.
†Temperature Derating: Plastic "P" Package: -12mW/°C from 65°C to 85°C
Ceramic "L" Package: -12mW/°C from 100°C to 125°C

MC14051B 8-Channel Analog Multiplexer/Demultiplexer
Controls: Inhibit 6, A 11, B 10, C 9
Switches In/Out: X0 13, X1 14, X2 15, X3 12, X4 1, X5 5, X6 2, X7 4 — Common Out/In X 3
V_{DD} = Pin 16, V_{SS} = Pin 8, V_{EE} = Pin 7

MC14052B Dual 4-Channel Analog Multiplexer/Demultiplexer
Controls: Inhibit 6, A 10, B 9
Switches In/Out: X0 12, X1 14, X2 15, X3 11, Y0 1, Y1 5, Y2 2, Y3 4 — Commons Out/In X 13, Y 3
V_{DD} = Pin 16, V_{SS} = Pin 8, V_{EE} = Pin 7

MC14053B Triple 2-Channel Analog Multiplexer/Demultiplexer
Controls: Inhibit 6, A 11, B 10, C 9
Switches In/Out: X0 12, X1 13, Y0 1, Y1 2, Z0 5, Z1 3 — Commons Out/In X 14, Y 15, Z 4
V_{DD} = Pin 16, V_{SS} = Pin 8, V_{EE} = Pin 7

Note: Control Inputs referenced to V_{SS}. Analog Inputs and Outputs reference to V_{EE}: V_{EE} must be ≤ V_{SS}.

ELECTRICAL CHARACTERISTICS

Characteristic	Symbol	V_{DD}	Test Conditions	Tlow* Min	Tlow* Max	25°C Min	25°C Typ#	25°C Max	Thigh* Min	Thigh* Max	Unit
SUPPLY REQUIREMENTS (Voltages Referenced to V_{EE})											
Power Supply Voltage Range	V_{DD}	—	$V_{DD} - 3 > V_{SS} > V_{EE}$	3	18	3	—	18	3	18	V
Quiescent Current Per Package (AL Device)	I_{DD}	5	Control Inputs V_{in} = V_{SS} or V_{DD}, Switch I/O: $V_{EE} < V_{iO} < V_{DD}$, and ΔV_{switch} ≤ 500 mV**	—	5	—	0.005	5	—	150	µA
		10		—	10	—	0.010	10	—	300	
		15		—	20	—	0.015	20	—	600	
Quiescent Current Per Package (CL/CP Device)	I_{DD}	5	Control Inputs V_{in} = V_{SS} or V_{DD}, Switch I/O: $V_{EE} < V_{iO} < V_{DD}$, and ΔV_{switch} ≤ 500 mV**	—	20	—	0.005	20	—	150	µA
		10		—	40	—	0.010	40	—	300	
		15		—	80	—	0.015	80	—	600	
Total Supply Current (Dynamic Plus Quiescent, Per Package)	$I_{D(AV)}$	5	T_A = 25°C only (The channel component, $(V_{in} - V_{out})/R_{on}$, is not included)	Typical (0.07 µA/kHz)f + I_{DD}							µA
		10		Typical (0.20 µA/kHz)f + I_{DD}							
		15		Typical (0.36 µA/kHz)f + I_{DD}							
CONTROL INPUTS — INHIBIT, A, B, C (Voltages Referenced to V_{SS})											
Low-Level Input Voltage	V_{IL}	5	R_{on} = per spec, I_{off} = per spec	—	1.5	—	2.25	1.5	—	1.5	V
		10		—	3.0	—	4.50	3.0	—	3.0	
		15		—	4.0	—	6.75	4.0	—	4.0	
High-Level Input Voltage	V_{IH}	5	R_{on} = per spec, I_{off} = per spec	3.5	—	3.5	2.75	—	3.5	—	V
		10		7.0	—	7.0	5.50	—	7.0	—	
		15		11.0	—	11.0	8.25	—	11.0	—	
Input Leakage Current (AL Device)	I_{in}	15	V_{in} = 0 or V_{DD}	—	±0.1	—	±0.00001	±0.1	—	±0.1	µA
Input Leakage Current (CL/CP Device)	I_{in}	15	V_{in} = 0 or V_{DD}	—	±0.3	—	±0.00001	±0.3	—	±0.3	µA
Input Capacitance	C_{in}	—	V_{in} = 0 or V_{DD}	—	—	—	5.0	7.5	—	—	pF
SWITCHES IN/OUT AND COMMONS OUT/IN — X, Y, Z (Voltages Referenced to V_{EE})											
Recommended Peak-to-Peak Voltage into or Out of the Switch	V_{iO}	—	Channel On or Off	0	V_{DD}	0	—	V_{DD}	0	V_{DD}	Vpp
Recommended Static or Dynamic Voltage Across the Switch (Figure 5)	ΔV_{switch}	—	Channel On	—	600	—	—	600	—	600	mV
Output Offset Voltage	V_{OO}	15	V_{in} = 0 V, No load	—	—	—	10	—	—	—	µV
ON Resistance (AL Device)	R_{on}	5	ΔV_{switch} ≤ 500 mV**, V_{in} = V_{IL} or V_{IH} (Control), and V_{in} = 0 to V_{DD} (Switch)	—	800	—	250	1050	—	1300	Ω
		10		—	400	—	120	500	—	550	
		15		—	220	—	80	280	—	320	
ON Resistance (CL/CP Device)	R_{on}	5	ΔV_{switch} ≤ 500 mV**, V_{in} = V_{IL} or V_{IH} (Control), and V_{in} = 0 to V_{DD} (Switch)	—	880	—	250	1050	—	1200	Ω
		10		—	450	—	120	500	—	520	
		15		—	250	—	80	280	—	300	
Δ ON Resistance Between Any Two Channels in the Same Package	ΔR_{on}	5	V_{in} = V_{IL} or V_{IH} (Control), and V_{in} = 0 to V_{DD} (Switch)	—	70	—	25	70	—	135	Ω
		10		—	50	—	10	50	—	95	
		15		—	45	—	10	45	—	65	
Off-Channel Leakage Current (AL Device) (Figure 10)	I_{off}	15	V_{in} = V_{IL} or V_{IH} (Control) Channel to Channel or Any One Channel	—	±100	—	±0.05	±100	—	±1000	nA
Off-Channel Leakage Current (CL/CP Device) (Figure 10)	I_{off}	15	V_{in} = V_{IL} or V_{IH} (Control) Channel to Channel or Any One Channel	—	±300	—	±0.05	±300	—	±1000	nA
Capacitance, Switch I/O	$C_{I/O}$	—	Inhibit = V_{DD}	—	—	—	10	—	—	—	pF
Capacitance, Common O/I	$C_{O/I}$	—	Inhibit = V_{DD} (MC14051B)	—	—	—	60	—	—	—	pF
			(MC14052B)	—	—	—	32	—	—	—	
			(MC14053B)	—	—	—	17	—	—	—	
Capacitance, Feedthrough (Channel Off)	$C_{I/O}$	—	Pins Not Adjacent	—	—	—	0.15	—	—	—	pF
			Pins Adjacent	—	—	—	0.47	—	—	—	

* T_{low} = -55°C for AL Device, -40°C for CL/CP Device.
T_{high} = +125°C for AL Device, +85°C for CL/CP Device.
Data labeled "Typ" is not to be used for design purposes, but is intended as an indication of the IC's potential performance.
**For voltage drops across the switch (ΔV_{switch}) >600 mV (>300 mV at high temperature), excessive V_{DD} current may be drawn; i.e. the current out of the switch may contain both V_{DD} and switch input components. The reliability of the device will be unaffected unless the Maximum Ratings are exceeded. (See first page of this data sheet.)

MOTOROLA

MC14532B

CMOS MSI
(LOW-POWER COMPLEMENTARY MOS)

8-BIT PRIORITY ENCODER

8-BIT PRIORITY ENCODER

The MC14532B is constructed with complementary MOS (CMOS) enhancement mode devices. The primary function of a priority encoder is to provide a binary address for the active input with the highest priority. Eight data inputs (D0 thru D7) and an enable input (E_{in}) are provided. Five outputs are available, three are address outputs (Q0 thru Q2), one group select (GS) and one enable output (E_{out}).

- Diode Protection on All Inputs
- Supply Voltage Range = 3.0 Vdc to 18 Vdc
- Capable of Driving Two Low-power TTL Loads or One Low-Power Schottky TTL Load over the Rated Temperature Range

L SUFFIX
CERAMIC PACKAGE
CASE 620

P SUFFIX
PLASTIC PACKAGE
CASE 648

ORDERING INFORMATION

A Series: -55°C to +125°C
MC14XXXBAL (Ceramic Package Only)

C Series: -40°C to +85°C
MC14XXXBCP (Plastic Package)
MC14XXXBCL (Ceramic Package)

PIN ASSIGNMENT

D4	1	16	V_{DD}
D5	2	15	E_{out}
D6	3	14	GS
D7	4	13	D3
E_{in}	5	12	D2
Q2	6	11	D1
Q1	7	10	D0
V_{SS}	8	9	Q0

MAXIMUM RATINGS* (Voltages Referenced to V_{SS})

Symbol	Parameter	Value	Unit
V_{DD}	DC Supply Voltage	-0.5 to +18.0	V
V_{in}, V_{out}	Input or Output Voltage (DC or Transient)	-0.5 to V_{DD} +0.5	V
I_{in}, I_{out}	Input or Output Current (DC or Transient), per Pin	±10	mA
P_D	Power Dissipation, per Package†	500	mW
T_{stg}	Storage Temperature	-65 to +150	°C
T_L	Lead Temperature (8-Second Soldering)	260	°C

*Maximum Ratings are those values beyond which damage to the device may occur.
†Temperature Derating: Plastic "P" Package: -12mW/°C from 65°C to 85°C
Ceramic "L" Package: -12mW/°C from 100°C to 125°C

TRUTH TABLE

INPUT										OUTPUT				
E_{in}	D7	D6	D5	D4	D3	D2	D1	D0	GS	Q2	Q1	Q0	E_{out}
0	X	X	X	X	X	X	X	X	0	0	0	0	1
1	X	X	X	X	X	X	X	1	1	1	1	1	0
1	X	X	X	X	X	X	1	0	1	1	1	0	0
1	X	X	X	X	X	1	0	0	1	1	0	1	0
1	X	X	X	X	1	0	0	0	1	1	0	0	0
1	X	X	X	1	0	0	0	0	1	0	1	1	0
1	X	X	1	0	0	0	0	0	1	0	1	0	0
1	X	1	0	0	0	0	0	0	1	0	0	1	0
1	1	0	0	0	0	0	0	0	1	0	0	0	0
1	0	0	0	0	0	0	0	0	0	0	0	0	1

X = Don't Care

This device contains protection circuitry to guard against damage due to high static voltages or electric fields. However, precautions must be taken to avoid applications of any voltage higher than maximum rated voltages to this high-impedance circuit. For proper operation, V_{in} and V_{out} should be constrained to the range $V_{SS} \leq (V_{in}$ or $V_{out}) \leq V_{DD}$.
Unused inputs must always be tied to an appropriate logic voltage level (e.g., either V_{SS} or V_{DD}). Unused outputs must be left open.

ELECTRICAL CHARACTERISTICS (Voltages Referenced to V_{SS})

Characteristic		Symbol	V_{DD} Vdc	T_{low}* Min	T_{low}* Max	25°C Min	25°C Typ #	25°C Max	T_{high}* Min	T_{high}* Max	Unit
Output Voltage "0" Level	$V_{in} = V_{DD}$ or 0	V_{OL}	5.0	—	0.05	—	0	0.05	—	0.05	Vdc
			10	—	0.05	—	0	0.05	—	0.05	
			15	—	0.05	—	0	0.05	—	0.05	
$V_{in} = 0$ or V_{DD} "1" Level		V_{OH}	5.0	4.95	—	4.95	5.0	—	4.95	—	Vdc
			10	9.95	—	9.95	10	—	9.95	—	
			15	14.95	—	14.95	15	—	14.95	—	
Input Voltage "0" Level		V_{IL}	5.0	—	1.5	—	2.25	1.5	—	1.5	Vdc
($V_O = 4.5$ or 0.5 Vdc)			10	—	3.0	—	4.50	3.0	—	3.0	
($V_O = 9.0$ or 1.0 Vdc)			15	—	4.0	—	6.75	4.0	—	4.0	
($V_O = 13.5$ or 1.5 Vdc)	"1" Level	V_{IH}	5.0	3.5	—	3.5	2.75	—	3.5	—	Vdc
($V_O = 0.5$ or 4.5 Vdc)			10	7.0	—	7.0	5.50	—	7.0	—	
($V_O = 1.0$ or 9.0 Vdc)			15	11.0	—	11.0	8.25	—	11.0	—	
($V_O = 1.5$ or 13.5 Vdc)											
Output Drive Current (AL Device) Source		I_{OH}	5.0	-3.0	—	-2.4	-4.2	—	-1.7	—	mAdc
($V_{OH} = 2.5$ Vdc)			5.0	-0.64	—	-0.51	-0.88	—	-0.36	—	
($V_{OH} = 4.6$ Vdc)			10	-1.6	—	-1.3	-2.25	—	-0.9	—	
($V_{OH} = 9.5$ Vdc)			15	-4.2	—	-3.4	-8.8	—	-2.4	—	
($V_{OH} = 13.5$ Vdc)											
($V_{OL} = 0.4$ Vdc) Sink		I_{OL}	5.0	0.64	—	0.51	0.88	—	0.36	—	mAdc
($V_{OL} = 0.5$ Vdc)			10	1.6	—	1.3	2.25	—	0.9	—	
($V_{OL} = 1.5$ Vdc)			15	4.2	—	3.4	8.8	—	2.4	—	
Output Drive Current (CL/CP Device) Source		I_{OH}	5.0	-2.5	—	-2.1	-2.4	—	-1.7	—	mAdc
($V_{OH} = 2.5$ Vdc)			5.0	-0.52	—	-0.44	-0.88	—	-0.36	—	
($V_{OH} = 4.6$ Vdc)			10	-1.3	—	-1.1	-2.25	—	-0.9	—	
($V_{OH} = 9.5$ Vdc)			15	-3.6	—	-3.0	-8.8	—	-2.4	—	
($V_{OH} = 13.5$ Vdc)											
($V_{OL} = 0.4$ Vdc) Sink		I_{OL}	5.0	0.52	—	0.44	0.88	—	0.36	—	mAdc
($V_{OL} = 0.5$ Vdc)			10	1.3	—	1.1	2.25	—	0.9	—	
($V_{OL} = 1.5$ Vdc)			15	3.6	—	3.0	8.8	—	2.4	—	
Input Current (AL Device)		I_{in}	15	—	±0.1	—	±0.00001	±0.1	—	±1.0	μAdc
Input Current (CL/CP Device)		I_{in}	15	—	±0.3	—	±0.00001	±0.3	—	±1.0	μAdc
Input Capacitance ($V_{in} = 0$)		C_{in}	—	—	—	—	5.0	7.5	—	—	pF
Quiescent Current (AL Device) (Per Package)		I_{DD}	5.0	—	5.0	—	0.005	5.0	—	150	μAdc
			10	—	10	—	0.010	10	—	300	
			15	—	20	—	0.015	20	—	600	
Quiescent Current (CL/CP Device) (Per Package)		I_{DD}	5.0	—	20	—	0.005	20	—	150	μAdc
			10	—	40	—	0.010	40	—	300	
			15	—	80	—	0.015	80	—	600	
Total Supply Current**† (Dynamic plus Quiescent, Per Package) ($C_L = 50$ pF on all outputs, all buffers switching)		I_T	5.0	$I_T = (1.74 \mu A/kHz) f + I_{DD}$							μAdc
			10	$I_T = (3.65 \mu A/kHz) f + I_{DD}$							
			15	$I_T = (5.73 \mu A/kHz) f + I_{DD}$							

$T_{low} = -55°C$ for AL Device, -40°C for CL/CP Device.
$T_{high} = +125°C$ for AL Device, +85°C for CL/CP Device.

#Data labelled "Typ" is not to be used for design purposes but is intended as an indication of the IC's potential performance.

**The formulas given are for the typical characteristics only at 25°C.

†To calculate total supply current at loads other than 50 pF:
$I_T(C_L) = I_T(50 \text{ pF}) + (C_L - 50) Vfk$

where: I_T is in μA (per package), C_L in pF, $V = (V_{DD} - V_{SS})$ in volts, f in kHz is input frequency, and $k = 0.005$.

MC14532B

255

National Semiconductor

Industrial Blocks

LM555/LM555C Timer

General Description

The LM555 is a highly stable device for generating accurate time delays or oscillation. Additional terminals are provided for triggering or resetting if desired. In the time delay mode of operation, the time is precisely controlled by one external resistor and capacitor. For astable operation as an oscillator, the free running frequency and duty cycle are accurately controlled with two external resistors and one capacitor. The circuit may be triggered and reset on falling waveforms, and the output circuit can source or sink up to 200 mA or drive TTL circuits.

Features

- Direct replacement for SE555/NE555
- Timing from microseconds through hours
- Operates in both astable and monostable modes
- Adjustable duty cycle
- Output can source or sink 200 mA
- Output and supply TTL compatible
- Temperature stability better than 0.005% per °C
- Normally on and normally off output

Applications

- Precision timing
- Pulse generation
- Sequential timing
- Time delay generation
- Pulse width modulation
- Pulse position modulation
- Linear ramp generator

Absolute Maximum Ratings

Supply Voltage	+18V
Power Dissipation (Note 1)	600 mW
Operating Temperature Ranges	
LM555C	0°C to +70°C
LM555	-55°C to +125°C
Storage Temperature Range	-65°C to +150°C
Lead Temperature (Soldering, 10 seconds)	300°C

Electrical Characteristics ($T_A = 25°C$, $V_{CC} = +5V$ to $+15V$, unless otherwise specified)

PARAMETER	CONDITIONS	LM555 MIN	LM555 TYP	LM555 MAX	LM555C MIN	LM555C TYP	LM555C MAX	UNITS
Supply Voltage		4.5		18	4.5		16	V
Supply Current	$V_{CC}=5V$, $R_L=\infty$		3	5		3	6	mA
	$V_{CC}=15V$, $R_L=\infty$ (Low State)(Note 2)		10	12		10	15	mA
Timing Error, Monostable								
Initial Accuracy	R_A, $R_B=1k$ to $100k$, $C=0.1\mu F$ (Note 3)		0.5			1		%
Drift with Temperature			30			50		ppm/°C
Accuracy over Temperature			1.5			1.5		%
Drift with Supply			0.05			0.1		%/V
Timing Error, Astable								
Initial Accuracy			1.5			2.25		%
Drift with Temperature			90			150		ppm/°C
Accuracy over Temperature			2.5			3.0		%
Drift with Supply			0.15			0.30		%/V
Threshold Voltage			0.667			0.667		x V_{CC}
Trigger Voltage	$V_{CC}=15V$	4.8	5	5.2		5		V
	$V_{CC}=5V$	1.45	1.67	1.9		1.67		V
Trigger Current			0.01	0.5		0.5	0.9	µA
Reset Voltage		0.4	0.5	1	0.4	0.5	1	V
Reset Current			0.1	0.4		0.1	0.4	mA
Threshold Current	(Note 4)		0.1	0.25		0.1	0.25	µA
Control Voltage Level	$V_{CC}=15V$	9.6	10	10.4	9	10	11	V
	$V_{CC}=5V$	2.9	3.33	3.8	2.6	3.33	4	V
Pin 7 Leakage Output High			1	100		1	100	nA
Pin 7 Set (Note 5)								
Output Low	$V_{CC}=15V$, $I_7=15$ mA		150			180		mV
Output Low	$V_{CC}=4.5V$, $I_7=4.5$ mA		70			80		mV
Output Voltage Drop (Low)	$V_{CC}=15V$							
	$I_{SINK}=10$ mA		0.1	0.15		0.1	0.25	V
	$I_{SINK}=50$ mA		0.4	0.5		0.4	0.75	V
	$I_{SINK}=100$ mA		2	2.2		2	2.5	V
	$I_{SINK}=200$ mA		2.5			2.5		V
	$V_{CC}=5V$							
	$I_{SINK}=8$ mA		0.1	0.25				V
	$I_{SINK}=5$ mA					0.25	0.35	V
Output Voltage Drop (High)	$I_{SOURCE}=200$ mA, $V_{CC}=15V$		12.5			12.5		V
	$I_{SOURCE}=100$ mA, $V_{CC}=15V$	13	13.3		12.75	13.3		V
	$V_{CC}=5V$	3	3.3		2.75	3.3		V
Rise Time of Output			100			100		ns
Fall Time of Output			100			100		ns

Note 1: For operating at elevated temperatures the device must be derated based on a +150°C maximum junction temperature and a thermal resistance of +45°C/W junction to case for TO-5 and +150°C/W junction to ambient for both packages.

Note 2: Supply current when output high typically 1 mA less at $V_{CC} = 5V$.

Note 3: Tested at $V_{CC} = 5V$ and $V_{CC} = 15V$.

Note 4: This will determine the maximum value of $R_A + R_B$ for 15V operation. The maximum total ($R_A + R_B$) is 20 MΩ.

Note 5: No protection against excessive pin 7 current is necessary providing the package dissipation rating will not be exceeded.

Schematic Diagram

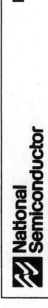

Connection Diagrams

Metal Can Package

TOP VIEW

Order Number LM555H, LM555CH
See NS Package H08C

Dual-In-Line Package

TOP VIEW

Order Number LM555CN
See NS Package N08B
Order Number LM555J or LM555CJ
See NS Package J08A

National Semiconductor

Operational Amplifiers/Buffers

LM741/LM741A/LM741C/LM741E Operational Amplifier

General Description

The LM741 series are general purpose operational amplifiers which feature improved performance over industry standards like the LM709. They are direct, plug-in replacements for the 709C, LM201, MC1439 and 748 in most applications.

The amplifiers offer many features which make their application nearly foolproof: overload pro-

tection on the input and output, no latch-up when the common mode range is exceeded, as well as freedom from oscillations.

The LM741C/LM741E are identical to the LM741/LM741A except that the LM741C/LM741E have their performance guaranteed over a 0°C to +70°C temperature range, instead of −55°C to +125°C.

Absolute Maximum Ratings

	LM741A	LM741E	LM741	LM741C
Supply Voltage	±22V	±22V	±22V	±18V
Power Dissipation (Note 1)	500 mW	500 mW	500 mW	500 mW
Differential Input Voltage	±30V	±30V	±30V	±30V
Input Voltage (Note 2)	±15V	±15V	±15V	±15V
Output Short Circuit Duration	Indefinite	Indefinite	Indefinite	Indefinite
Operating Temperature Range	−55°C to +125°C	0°C to +70°C	−55°C to +125°C	0°C to +70°C
Storage Temperature Range	−65°C to +150°C	−65°C to +150°C	−65°C to +150°C	−65°C to +150°C
Lead Temperature (Soldering, 10 seconds)	300°C	300°C	300°C	300°C

Electrical Characteristics (Note 3)

PARAMETER	CONDITIONS	LM741A/LM741E MIN	TYP	MAX	LM741 MIN	TYP	MAX	LM741C MIN	TYP	MAX	UNITS
Input Offset Voltage	$T_A = 25°C$										
	$R_S \leq 10\ k\Omega$					1.0	5.0		2.0	6.0	mV
	$R_S \leq 50\Omega$		0.8	3.0							mV
	$T_{AMIN} \leq T_A \leq T_{AMAX}$										
	$R_S \leq 50\Omega$			4.0							mV
	$R_S \leq 10\ k\Omega$						6.0			7.5	mV
Average Input Offset Voltage Drift				15							$\mu V/°C$
Input Offset Voltage Adjustment Range	$T_A = 25°C$, $V_S = \pm20V$	±10				±15			±15		mV
Input Offset Current	$T_A = 25°C$		3.0	30		20	200		20	200	nA
	$T_{AMIN} \leq T_A \leq T_{AMAX}$			70		85	500			300	nA
Average Input Offset Current Drift				0.5							$nA/°C$
Input Bias Current	$T_A = 25°C$		30	80		80	500		80	500	nA
	$T_{AMIN} \leq T_A \leq T_{AMAX}$			0.210			1.5			0.8	μA
Input Resistance	$T_A = 25°C$, $V_S = \pm20V$	1.0	6.0		0.3	2.0		0.3	2.0		$M\Omega$
	$T_{AMIN} \leq T_A \leq T_{AMAX}$, $V_S = \pm20V$	0.5									$M\Omega$
Input Voltage Range	$T_A = 25°C$				±12	±13		±12	±13		V
	$T_{AMIN} \leq T_A \leq T_{AMAX}$										V
Large Signal Voltage Gain	$T_A = 25°C$, $R_L \geq 2\ k\Omega$										
	$V_S = \pm20V$, $V_O = \pm15V$	50									V/mV
	$V_S = \pm15V$, $V_O = \pm10V$				50	200		20	200		V/mV
	$T_{AMIN} \leq T_A \leq T_{AMAX}$, $R_L \geq 2\ k\Omega$										
	$V_S = \pm20V$, $V_O = \pm15V$	32									V/mV
	$V_S = \pm15V$, $V_O = \pm10V$				25			15			V/mV
	$V_S = \pm5V$, $V_O = \pm2V$	10									V/mV
Output Voltage Swing	$V_S = \pm20V$										
	$R_L \geq 10\ k\Omega$	±16									V
	$R_L \geq 2\ k\Omega$	±15									V
	$V_S = \pm15V$				±12	±14		±12	±14		V
	$R_L \geq 10\ k\Omega$				±10	±13		±10	±13		V
	$R_L \geq 2\ k\Omega$										V
Output Short Circuit Current	$T_A = 25°C$	10	25	35		25			25		mA
	$T_{AMIN} \leq T_A \leq T_{AMAX}$	10		40							mA
Common-Mode Rejection Ratio	$T_{AMIN} \leq T_A \leq T_{AMAX}$										
	$R_S \leq 10\ k\Omega$, $V_{CM} = \pm12V$				70	90		70	90		dB
	$R_S \leq 50\ k\Omega$, $V_{CM} = \pm12V$	80	95								dB

Schematic and Connection Diagrams (Top Views)

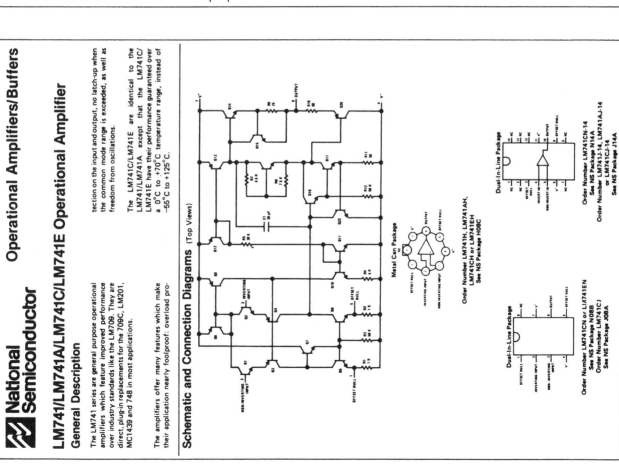

Metal Can Package

Order Number LM741H, LM741AH, LM741CH or LM741EH
See NS Package H08C

Dual-In-Line Package

Order Number LM741CN or LM741EN
See NS Package N08B
Order Number LM741CJ
See NS Package J08A

Dual-In-Line Package

Order Number LM741CN-14
See NS Package N14A
Order Number LM741J-14, LM741AJ-14
or LM741CJ-14
See NS Package J14A

ADC0801/ADC0802/ADC0803/ADC0804/ADC0805
8-Bit μP Compatible A/D Converters

General Description

The ADC0801, ADC0802, ADC0803, ADC0804 and ADC0805 are CMOS 8-bit successive approximation A/D converters that use a differential potentiometric ladder—similar to the 256R products. These converters are designed to allow operation with the NSC800 and INS8080A derivative control bus with TRI-STATE® output latches directly driving the data bus. These A/Ds appear like memory locations or I/O ports to the microprocessor and no interfacing logic is needed.

Differential analog voltage inputs allow increasing the common-mode rejection and offsetting the analog zero input voltage value. In addition, the voltage reference input can be adjusted to allow encoding any smaller analog voltage span to the full 8 bits of resolution.

Features

- Compatible with 8080 μP derivatives—no interfacing logic needed - access time - 135 ns
- Easy interface to all microprocessors, or operates "stand alone"
- Differential analog voltage inputs
- Logic inputs and outputs meet both MOS and TTL voltage level specifications
- Works with 2.5V (LM336) voltage reference
- On-chip clock generator
- 0V to 5V analog input voltage range with single 5V supply
- No zero adjust required
- 0.3" standard width 20-pin DIP package
- 20-pin molded chip carrier or small outline package
- Operates ratiometrically or with 5 V$_{DC}$, 2.5 V$_{DC}$, or analog span adjusted voltage reference

Key Specifications

- Resolution   8 bits
- Total error   ±¼ LSB, ±½ LSB and ±1 LSB
- Conversion time   100 μs

Typical Applications

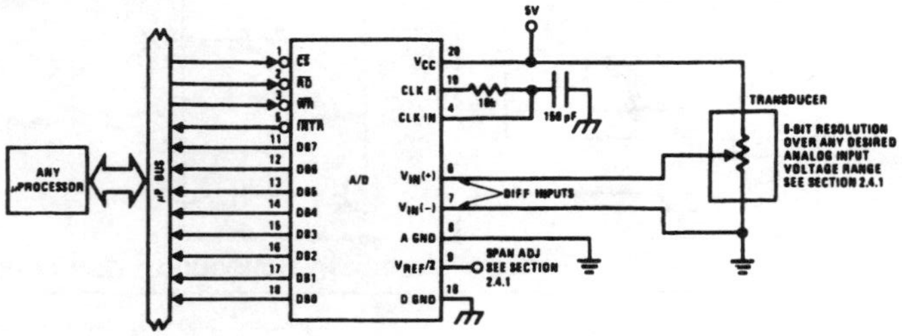

TL/H/5671-1

8080 Interface

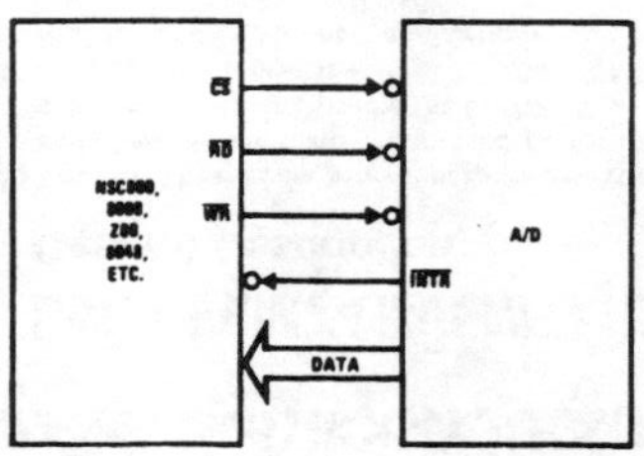

TL/H/5671-31

Error Specification (Includes Full-Scale, Zero Error, and Non-Linearity)

Part Number	Full-Scale Adjusted	V$_{REF}$/2 = 2.500 V$_{DC}$ (No Adjustments)	V$_{REF}$/2 = No Connection (No Adjustments)
ADC0801	±¼ LSB		
ADC0802		±½ LSB	
ADC0803	±½ LSB		
ADC0804		±1 LSB	
ADC0805			±1 LSB

Absolute Maximum Ratings (Notes 1 & 2)

If Military/Aerospace specified devices are required, please contact the National Semiconductor Sales Office/Distributors for availability and specifications.

Supply Voltage (V$_{CC}$) (Note 3)	6.5V
Voltage	
Logic Control Inputs	−0.3V to +18V
At Other Input and Outputs	−0.3V to (V$_{CC}$ +0.3V)
Lead Temp. (Soldering, 10 seconds)	
Dual-In-Line Package (plastic)	260°C
Dual-In-Line Package (ceramic)	300°C
Surface Mount Package	
Vapor Phase (60 seconds)	215°C
Infrared (15 seconds)	220°C
Storage Temperature Range	−65°C to +150°C
Package Dissipation at T$_A$=25°C	875 mW
ESD Susceptibility (Note 10)	800V

Operating Ratings (Notes 1 & 2)

Temperature Range	T$_{MIN}$≤T$_A$≤T$_{MAX}$
ADC0801/02LJ	−55°C≤T$_A$≤+125°C
ADC0801/02/03/04LCJ	−40°C≤T$_A$≤+85°C
ADC0801/02/03/05LCN	−40°C≤T$_A$≤+85°C
ADC0804LCN	0°C≤T$_A$≤+70°C
ADC0802/03/04LCV	0°C≤T$_A$≤+70°C
ADC0802/03/04LCWM	0°C≤T$_A$≤+70°C
Range of V$_{CC}$	4.5 V$_{DC}$ to 6.3 V$_{DC}$

Electrical Characteristics

The following specifications apply for V$_{CC}$ = 5 V$_{DC}$, T$_{MIN}$≤T$_A$≤T$_{MAX}$ and f$_{CLK}$ = 640 kHz unless otherwise specified.

Parameter	Conditions	Min	Typ	Max	Units
ADC0801: Total Adjusted Error (Note 8)	With Full-Scale Adj. (See Section 2.5.2)			±¼	LSB
ADC0802: Total Unadjusted Error (Note 8)	V$_{REF}$/2 = 2.500 V$_{DC}$			±½	LSB
ADC0803: Total Adjusted Error (Note 8)	With Full-Scale Adj. (See Section 2.5.2)			±½	LSB
ADC0804: Total Unadjusted Error (Note 8)	V$_{REF}$/2 = 2.500 V$_{DC}$			±1	LSB
ADC0805: Total Unadjusted Error (Note 8)	V$_{REF}$/2-No Connection			±1	LSB
V$_{REF}$/2 Input Resistance (Pin 9)	ADC0801/02/03/05	2.5	8.0		kΩ
	ADC0804 (Note 9)	0.75	1.1		kΩ
Analog Input Voltage Range	(Note 4) V(+) or V(−)	Gnd−0.05		V$_{CC}$ +0.05	V$_{DC}$
DC Common-Mode Error	Over Analog Input Voltage Range		±1/16	±⅛	LSB
Power Supply Sensitivity	V$_{CC}$ = 5 V$_{DC}$ ±10% Over Allowed V$_{IN}$(+) and V$_{IN}$(−) Voltage Range (Note 4)		±1/16	±⅛	LSB

AC Electrical Characteristics

The following specifications apply for V$_{CC}$ = 5 V$_{DC}$ and T$_A$ = 25°C unless otherwise specified.

Symbol	Parameter	Conditions	Min	Typ	Max	Units
T$_C$	Conversion Time	f$_{CLK}$ = 640 kHz (Note 6)	103		114	μs
T$_C$	Conversion Time	(Note 5, 6)	66		73	1/f$_{CLK}$
f$_{CLK}$	Clock Frequency	V$_{CC}$ = 5V, (Note 5)	100	640	1460	kHz
	Clock Duty Cycle	(Note 5)	40		60	%
CR	Conversion Rate in Free-Running Mode	INTR tied to WR with CS = 0 V$_{DC}$, f$_{CLK}$ = 640 kHz	8770		9708	conv/s
t$_W$(WR)L	Width of WR Input (Start Pulse Width)	CS = 0 V$_{DC}$ (Note 7)	100			ns
t$_{ACC}$	Access Time (Delay from Falling Edge of RD to Output Data Valid)	C$_L$ = 100 pF		135	200	ns
t$_{1H}$, t$_{0H}$	TRI-STATE Control (Delay from Rising Edge of RD to Hi-Z State)	C$_L$ = 10 pF, R$_L$ = 10k (See TRI-STATE Test Circuits)		125	200	ns
t$_{WI}$, t$_{RI}$	Delay from Falling Edge of WR or RD to Reset of INTR			300	450	ns
C$_{IN}$	Input Capacitance of Logic Control Inputs			5	7.5	pF
C$_{OUT}$	TRI-STATE Output Capacitance (Data Buffers)			5	7.5	pF
CONTROL INPUTS [Note: CLK IN (Pin 4) is the input of a Schmitt trigger circuit and is therefore specified separately]						
V$_{IN}$ (1)	Logical "1" Input Voltage (Except Pin 4 CLK IN)	V$_{CC}$ = 5.25 V$_{DC}$	2.0		15	V$_{DC}$

(M) MOTOROLA

MC1408 MC1508

EIGHT-BIT MULTIPLYING DIGITAL-TO-ANALOG CONVERTER

SILICON MONOLITHIC INTEGRATED CIRCUIT

L SUFFIX
CERAMIC PACKAGE
CASE 620-02

P SUFFIX
PLASTIC PACKAGE
CASE 648-05

Specifications and Applications Information

EIGHT-BIT MULTIPLYING DIGITAL-TO-ANALOG CONVERTER

. . . designed for use where the output current is a linear product of an eight-bit digital word and an analog input voltage.

- Eight-Bit Accuracy Available in Both Temperature Ranges
  Relative Accuracy: ±0.19% Error maximum (MC1408L8, MC1408P8, MC1508L8)
- Seven and Six-Bit Accuracy Available with MC1408 Designated by 7 or 6 Suffix after Package Suffix
- Fast Settling Time — 300 ns typical
- Noninverting Digital Inputs are MTTL and CMOS Compatible
- Output Voltage Swing — +0.4 V to -5.0 V
- High-Speed Multiplying Input Slew Rate 4.0 mA/µs
- Standard Supply Voltages: +5.0 V and -5.0 V to -15 V

FIGURE 1 — D-to-A TRANSFER CHARACTERISTICS

I_O, OUTPUT CURRENT (mA)

INPUT DIGITAL WORD

(00000000)   (11111111)

FIGURE 2 — BLOCK DIAGRAM

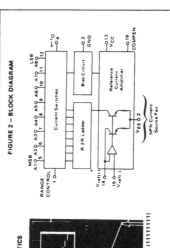

TYPICAL APPLICATIONS

- Tracking A-to-D Converters
- Successive Approximation A-to-D Converters
- 2 1/2 Digit Panel Meters and DVM's
- Waveform Synthesis
- Sample and Hold
- Peak Detector
- Programmable Gain and Attenuation
- CRT Character Generation
- Audio Digitizing and Decoding
- Programmable Power Supplies
- Analog-Digital Multiplication
- Digital-Digital Multiplication
- Analog-Digital Division
- Digital Addition and Subtraction
- Speech Compression and Expansion
- Stepping Motor Drive

MAXIMUM RATINGS (T_A = +25°C unless otherwise noted.)

Rating	Symbol	Value	Unit
Power Supply Voltage	V_{CC} / V_{EE}	+5.5 / -16.5	Vdc
Digital Input Voltage	V_5 thru V_{12}	0 to +5.5	Vdc
Applied Output Voltage	V_O	+0.5, -5.2	Vdc
Reference Current	I_{14}	5.0	mA
Reference Amplifier Inputs	V_{14}, V_{15}	V_{CC}, V_{EE}	Vdc
Operating Temperature Range MC1508 MC1408 Series	T_A	-55 to +125 0 to +75	°C
Storage Temperature Range	T_{stg}	-65 to +150	°C

ELECTRICAL CHARACTERISTICS (V_{CC} = +5.0 Vdc, V_{EE} = -15 Vdc, $V_{ref}/R14$ = 2.0 mA, MC1508L8: T_A = -55°C to +125°C. MC1408L Series: T_A = 0 to +75°C unless otherwise noted. All digital inputs at high logic level.)

Characteristic	Figure	Symbol	Min	Typ	Max	Unit
Relative Accuracy (Error relative to full scale I_O)	4	E_r				%
MC1508L8, MC1408L8, MC1408P8			—	—	±0.19	
MC1408P7, MC1408L7, See Note 1			—	—	±0.39	
MC1408P6, MC1408L6, See Note 1			—	—	±0.78	
Settling Time to within 1/2 LSB (includes t_{PLH})(T_A = +25°C)See Note 2	5	t_S	—	300	—	ns
Propagation Delay Time T_A = +25°C	5	t_{PLH}, t_{PHL}	—	30	100	ns
Output Full Scale Current Drift		TC_{IO}	—	-20	—	PPM/°C
Digital Input Logic Levels (MSB)	3					Vdc
High Level, Logic "1"		V_{IH}	2.0	—	—	
Low Level, Logic "0"		V_{IL}	—	—	0.8	
Digital Input Current (MSB)	3					mA
High Level, V_{IH} = 5.0 V		I_{IH}		0	0.04	
Low Level, V_{IL} = 0.8 V		I_{IL}		-0.4	-0.8	
Reference Input Bias Current (Pin 15)	3	I_{15}		-1.0	-5.0	µA
Output Current Range	3	I_{OR}				mA
V_{EE} = -5.0 V			0	2.0	2.1	
V_{EE} = -15 V, T_A = 25°C			0	2.0	4.2	
Output Current V_{ref} = 2.000 V, R14 = 1000 Ω	3	I_O	1.9	1.99	2.1	mA
Output Current (All bits low)	3	$I_{O(min)}$		0	4.0	µA
Output Voltage Compliance (E_r ≤ 0.19% at T_A = +25°C)	3	V_O			-0.55, +0.4 / -5.0, -0.4	Vdc
Pin 1 grounded / Pin 1 open, V_{EE} below -10 V						
Reference Current Slew Rate	6	SR I_{ref}		4.0		mA/µs
Output Current Power Supply Sensitivity		PSRRI(-)		0.5	2.7	µA/V
Power Supply Current (All bits low)	3	I_{CC} / I_{EE}		+13.5 / -7.5	+22 / -13	mA
Power Supply Voltage Range (T_A = +25°C)	3	V_{CCR} / V_{EER}	+4.5 / -4.5	+5.0 / -15	+5.5 / -16.5	Vdc
Power Dissipation	3	P_D				mW
All bits low V_{EE} = -5.0 Vdc				105	170	
V_{EE} = -15 Vdc				190	305	
All bits high V_{EE} = -5.0 Vdc				90	—	
V_{EE} = -15 Vdc				160	—	

Note 1. All current switches are tested to guarantee at least 50% of rated output current.
Note 2. All bits switched.

MOTOROLA LINEAR/INTERFACE DEVICES

Appendix B

List of Materials for the Experiments

Integrated Circuits:

TTL	*TTL*
7400	7492A
7402	7493A
7404	74121
7408	74LS139A
7432	74151A
7447A	74LS153
7474	74175
74LS76A	74LS189
7483A	74191
7485	74195
7486	

CMOS	*Linear*
4069	LM555
4071	LM741
4081	ADC0804
14051B	MC1408
14532B	

Display:
MAN-72 (or equivalent)

Resistors:
An assortment of ¼-W resistors is suggested. The most widely used values are 330 Ω, 390 Ω, 1.0 kΩ, 2 kΩ, 10 kΩ, and 100 kΩ.

Capacitors:
One 0.01 µF
Three 0.1 µF
One 1.0 µF
One 100 µF

Miscellaneous:
One 1 kΩ potentiometer
LEDs
Signal diodes (1N914 or equivalent)
4-position DIP switch
Two SPST N.O. pushbuttons
Two CdS photocells (Jameco 120299 or equivalent)
Solderless breadboard (Radio Shack #276-174 or equivalent)